OPTION PRICING IN
INCOMPLETE MARKETS

Modeling Based on Geometric Lévy Processes and
Minimal Entropy Martingale Measures

Series in Quantitative Finance ISSN: 1756-1604

Series in Quantitative Finance – Vol. 3

OPTION PRICING IN INCOMPLETE MARKETS

Modeling Based on Geometric Lévy Processes and
Minimal Entropy Martingale Measures

Yoshio Miyahara

Nagoya City University, Japan

Imperial College Press

Published by

Imperial College Press
57 Shelton Street
Covent Garden
London WC2H 9HE

Distributed by

World Scientific Publishing Co. Pte. Ltd.
5 Toh Tuck Link, Singapore 596224
USA office: 27 Warren Street, Suite 401-402, Hackensack, NJ 07601
UK office: 57 Shelton Street, Covent Garden, London WC2H 9HE

British Library Cataloguing-in-Publication Data
A catalogue record for this book is available from the British Library.

Series in Qualitative Finance — Vol. 3
OPTION PRICING IN INCOMPLETE MARKETS
Modeling Based on Geometric Lévy Processes and Minimal Entropy Martingale Measures

Copyright © 2012 by Imperial College Press

For photocopying of material in this volume, please pay a copying fee through the Copyright Clearance Center, Inc., 222 Rosewood Drive, Danvers, MA 01923, USA. In this case permission to photocopy is not required from the publisher.

ISBN-13 978-1-84816-347-8
ISBN-10 1-84816-347-9

Printed in Singapore by B & Jo Enterprise Pte Ltd

Preface

Mathematical finance theory has been established as one sector of finance theory. This theory is based on probability theory and many probabilists have contributed to this field.

The Black–Scholes model is a typical model for the complete market. This model is an outstanding model convenient to analyze, but it is well-known that in the real world the completeness of the market is not usually satisfied. The distribution of the log return of an asset usually has a fat tail and asymmetry, and the market is usually incomplete. Therefore we need other models for the incomplete market.

The *geometric Lévy process* (GLP) model is one of the most important models for the incomplete market. This model is able to possess a fat tail property, an asymmetric distribution and smile/smirk properties of implied volatility.

An incomplete market has many martingale measures (or risk-neutral measures) by the second fundamental theorem of mathematical finance. So we have to select a suitable martingale measure among them in order to discuss option pricing based on arbitrage theory. Many kinds of martingale measures have been proposed for this. Among them the minimal entropy martingale measure (MEMM) is the most important candidate.

Among the many GLP models, the importance of the *geometric stable process* (GSP) model is recognized by Fama [36] and Mandelbrot [77]. Since then many researchers have studied this model. The suitable martingale measure for the GSP model had not been evident for many years, but since the existence of the MEMM for GSP model was proved and an explicit form of it was obtained, the GSP model can be applied to a wide class of problems concerning option pricing.

The main subject of this book is the description of the [GLP & MEMM] (geometric Lévy process and minimal entropy martingale measure) option pricing models. We introduce these models and explain how to apply them to practical problems. This will be covered in Chapters 7, 8, and 9.

Before we introduce the [GLP & MEMM] option pricing models, we will explain the underlying ideas needed to understand our models. In Chapter 1 we give basic concepts in mathematical finance theory. In Chapter 2 Lévy processes and GLP models are introduced and described briefly. The following four chapters (from Chapter 3 to Chapter 6) are devoted to the investigation of martingale measures. In those chapters we stress the importance of Esscher-transformed martingale measures and minimal distance martingale measures. We also show that the MEMM is a special martingale measure among the set of all martingale measures in the sense that it is the minimal distance martingale measure determined by the entropy distance, and it is also an Esscher-transformed martingale measure determined by the simple risk process.

The importance of the [GLP & MEMM] option pricing models will be explained in Chapters 7 and 8. Furthermore it is explained in Chapter 9 that the [GSP & MEMM] model should be the most important one.

In this book we mainly treat a one-dimensional case. But many of the results for one-dimensional cases can be extended to multi-dimensional cases. In Chapter 10 we study a multi-dimensional case. The results obtained there are not simple extensions of the one-dimensional case and contain new concepts, new problems, and new ideas which are inherent to the multi-dimensional case. These days, a theory for non-arbitrage-free markets is needed. The idea of risk measure or value measure is an attempt to construct this theory. We introduce the idea of a risk-sensitive value measure and apply it to the problems of portfolio evaluation.

Appendix A is devoted to the explanation of the generalized method of moments for the parameter estimation of Lévy processes. The estimation problem of Lévy processes is related to calibration problems.

I hope that this book will be helpful to researchers, students, practitioners, and engineers of mathematical finance or risk management. This book is written for academics with a solid knowledge in "classical financial mathematics" and also for advanced practitioners who seek an introduction to the topic of option pricing in incomplete markets and into the recent Lévy process modeling. The necessary preliminary knowledge to read the book is briefly stated in Chapters 1 and 2. If a reader has a difficulty in following these chapters, I recommend first studying some of the books referred to in the Notes of Chapters 1 and 2.

I have been studying the [GLP & MEMM] models for more than a decade. During this period I have collaborated with Alexander Novikov, Tsukasa Fujiwara, Monique Jeanblanc, Susanne Klöppel, Tetsuya Misawa, Yoshiki Tsujii, and Naruhiko Moriwaki. Almost all of the results that were obtained through joint works with these people are contained in this book. Without these joint works this book could not have been published. I thank these colleagues from the bottom of my heart.

Yoshio Miyahara
Nagoya 2011

Contents

Chapter 1

Basic Concepts in Mathematical Finance

In this chapter, we give an overview of basic concepts in mathematical finance theory, and then explain those concepts in very simple cases, namely in the single-term finite market.

1.1 Price Processes

Price processes of financial assets are usually modeled as stochastic processes. So mathematical finance theory is based on probability theory, particularly on the theory of stochastic processes.

The price process of an underlying asset is generally denoted by S_t in this book. The process S_t is usually assumed to be positive and is expressed in the following form

$$S_t = e^{Z_t}. \tag{1.1}$$

The time parameter t usually runs in $[0, T]$, $T > 0$, or $[0, \infty)$, and sometimes we consider the discrete time case ($t = 0, 1, \ldots, T$).

1.2 No-arbitrage and Martingale Measures

In mathematical finance theory, properties of the market where the financial assets are traded are vitally important. If the market works well, then the economy should work well, but if the market does not work well, then the economy shouldn't work.

An important property of the market is its efficiency. This is the "no-arbitrage" or "no free lunch" assumption in mathematical finance theory. A brief definition of arbitrage is: "an arbitrage opportunity is the possibility to make a profit in a financial market without risk and without net investment of capital" (see Delbaen and Schachermayer [30] p.4). The no-arbitrage assumption means that a market does not allow any arbitrage opportunity.

The theory which is constructed on the no-arbitrage assumption is called "arbitrage theory". In arbitrage theory, the martingale measure plays an essential role.

1.3 Complete and Incomplete Markets

If the market has enough commodities, then a new commodity should be a replica of old ones, and we don't need other new commodities. This concept of sufficient commodities in the market is the meaning of completeness in the market.

1.4 Fundamental Theorems

The two concepts introduced above are characterized by the concept of the martingale measure. The following two theorems are well known. (See Delbaen and Schachermayer (2006) [30] or Björk (2004) [9] for details.)

Theorem 1.1. (First Fundamental Theorem in Mathematical Finance). A necessary and sufficient condition for the absence of arbitrage opportunities is the existence of the martingale measure of the underlying asset process.

Theorem 1.2. (Second Fundamental Theorem in Mathematical Finance). Assume the absence of arbitrage opportunities. Then a necessary and sufficient condition for the completeness of the market is the uniqueness of the martingale measure.

If the market is arbitrage-free and complete, then the price of a contingent claim B, $\pi(B)$, is determined by

$$\pi(B) = E_Q[e^{-rT}B], \qquad (1.2)$$

where Q is the unique martingale measure and r is the interest rate of the bond. In the case where the market satisfies the no-arbitrage assumption but does not satisfy the completeness assumption, then the price $\pi(B)$ is supposed to be in the following interval:

$$\pi(B) \in \left[\inf_{Q \in \mathcal{M}} E_Q[e^{-rT}B], \ \sup_{Q \in \mathcal{M}} E_Q[e^{-rT}B] \right], \tag{1.3}$$

where $\mathcal{M}$ is the set of all equivalent martingale measures. (See Theorem 2.4.1 in Delbaen and Schachermayer [30].)

1.5 The Black–Scholes Model

The most popular and fundamental model in mathematical finance is the Black–Scholes model (geometric Brownian motion model). The explicit form of the underlying asset process of this model is given by

$$S_t = S_0 e^{(\mu - \frac{1}{2}\sigma^2)t + \sigma W_t}, \tag{1.4}$$

or equivalently in the stochastic differential equation (SDE) form

$$dS_t = S_t \left(\mu dt + \sigma dW_t \right), \tag{1.5}$$

where μ is a real number, σ is a positive real number, and W_t is a Wiener process (standard Brownian motion).

The risk-neutral measure ($=$ martingale measure) Q is uniquely determined by Girsanov's theorem. Under Q the process $\tilde{W}_t = W_t + (\mu - r)\sigma^{-1}t$ is a Wiener process and the price process S_t is expressed in the form

$$S_t = S_0 e^{(r - \frac{1}{2}\sigma^2)t + \sigma \tilde{W}_t} \quad \text{or} \quad dS_t = S_t \left(rdt + \sigma d\tilde{W}_t \right), \tag{1.6}$$

where r is the constant interest rate of a risk-free asset. The price of an option X is given by $e^{-rT}E_Q[X]$. The theoretical Black–Scholes price of the European call option, $C(S_0, K, T)$, with the strike price K and the fixed maturity T is given by the following formula:

$$C_K = C(S_0, K, T) = e^{-rT}E_Q[(S_T - K)^+] = S_0 N(d_1) - e^{-rT}KN(d_2), \tag{1.7}$$

where $N(d)$ is the normal distribution function and

$$d_1 = \frac{\log \frac{S_0}{K} + (r + \frac{\sigma^2}{2})T}{\sigma\sqrt{T}}, \quad d_2 = \frac{\log \frac{S_0}{K} + (r - \frac{\sigma^2}{2})T}{\sigma\sqrt{T}} = d_1 - \sigma\sqrt{T}. \tag{1.8}$$

1.6   Properties of the Black–Scholes Model

We summarize the basic properties of the Black–Scholes model as follows.

1.6.1   *Distribution of log returns*

The log return is the increment of the logarithm of S_t,

$$\triangle \log S_t = \log S_{t+\triangle t} - \log S_t = (\mu - \frac{1}{2}\sigma^2)\triangle t + \sigma \triangle W_t, \qquad (1.9)$$

and the log return process is $(\mu - \frac{1}{2}\sigma^2)t + \sigma W_t$.

The distribution of the log return (or the log return process) of the Black–Scholes model is normal. This is convenient for the calculation of the option prices. For example, we have obtained the explicit formula of the price of European call options. However, it is said that the distributions of the log returns in the real market usually have a fat tail and asymmetry. These facts suggest the necessity of considering another model.

1.6.2   *Historical volatility and implied volatility*

Under the setting of the Black–Scholes model, the historical volatility of the process is defined as the estimated value of σ based on the sequential data of the price process S_t. We denote it by $\widehat{\sigma}$. On the other hand, the implied volatility is defined in the following way. Suppose that the market price of the European call option with the strike K, say $C_K^{(m)}$ were given, then the value of σ which satisfies the equation

$$S_0 N(d_1) - e^{-rT} K N(d_2) = C_K^{(m)} \qquad (1.10)$$

is the implied volatility, and this value is denoted by $\sigma_K^{(im)}$. It should be noted that the implied volatility $\sigma_K^{(im)}$ depends on the strike value K but, on the contrary, that the historical volatility $\widehat{\sigma}$ does not depend on K.

We first consider the case where the market value of options obeys the Black–Scholes model, and so the market price $C_K^{(m)}$ is equal to the theoretical Black–Scholes price C_K. In this case the solution of the equation for the implied volatility is equal to the original σ and it holds true that $\sigma_K^{(im)} = \sigma = $ constant. This means that if the market exactly obeys the Black–Scholes model, then the implied volatility $\sigma_K^{(im)}$ should be equal to the historical volatility $\widehat{\sigma}$.

But in the real world this is not true. It is well known that the implied volatility is not equal to the historical volatility, and the implied volatility

$\sigma_K^{(im)}$ is sometimes a convex function of K, and sometimes a combination of a convex part and a concave part. These properties are the so-called "volatility smile or smirk" properties.

1.7 Generalization of the Black–Scholes Model

The Black–Scholes model is a complete market model, but it is said that the real market is incomplete in general. So we have to construct a suitable model for the incomplete market.

1.7.1 *Geometric Lévy process models*

We start from the explicit form of geometric Brownian motion:

$$S_t = S_0 e^{(\mu - \frac{1}{2}\sigma^2)t + \sigma W_t}. \tag{1.11}$$

It is a natural idea to replace the Wiener process with a more general Lévy process Z_t and consider the process

$$S_t = S_0 e^{Z_t}. \tag{1.12}$$

The processes of this type are called the geometric Lévy processes (GLP) or exponential Lévy processes. The [GLP & MEMM] pricing models, which are explained in Chapter 7, are of this type of generalization of Black–Scholes model.

The class of Lévy processes is very diverse and the distributions of S_t may have a at tail and may be asymmetric, and the geometric Lévy process models are generally incomplete market models. These models are studied in Chapter 2.

1.7.2 *Stochastic volatility models*

We start from the SDE form of the Black–Scholes model,

$$dS_t = S_t (\mu dt + \sigma dW_t). \tag{1.13}$$

When we randomize the volatility σ and consider the equation

$$dS_t = S_t (\mu dt + \tilde{\sigma}_t dW_t), \tag{1.14}$$

where $\tilde{\sigma}_t$ is a stochastic process, then we obtain the so-called stochastic volatility model.

This model is a very natural one when we think the volatility is dependent on the economic environment. (See Chapter 15 of Cont and Tankov
(2004) [25] for this model.)

Notes

For a general introduction to mathematical finance theory, see the following books:

Björk, T. (2004) [9],
Föllmer, H. and Shied, A. (2004) [38],
Jeanblanc, M., Yor, M. and Chesney, M. (2009) [59],
Karatzas, I. and Shreve, S.E. (1998) [64],
Pliska, R.S. (1997) [101],
Shiryaev, A.N. (1999) [116],
Shreve, S.E. (2003) [117],
Shreve, S.E. (2004) [118].

For a study on elementary probability theory there are many books, for
example:

Feller, W. (1966) [37],
Williams, D. (1991) [120].

Chapter 2

Lévy Processes and Geometric Lévy Process Models

In this chapter we summarize elementary facts on Lévy processes, and we introduce the geometric Lévy process model. We will now consider a one-dimensional case; a multi-dimensional case is treated only in Chapter 10.

2.1 Lévy Processes

A comprehensive text book on Lévy processes is Sato, K., "Lévy Processes and Infinitely Divisible Distribution", Cambridge University Press (1999). We recommend this book for further details.

2.1.1 *Definitions and properties*

We define Lévy processes as follows.

Definition 2.1. Suppose that a probability space (Ω, F, P) and a filtration $\{\mathcal{F}_t, t \geq 0\}$ are given. A continuous time stochastic process $\{Z_t, 0 \leq t \leq T\}$ defined on the probability space (Ω, F, P) is a Lévy process if the following conditions are satisfied.

(1) (independent increments property) For any $0 \leq t_0 < t_1 < \cdots < t_n \leq T$, $Z_{t_1} - Z_{t_0}, Z_{t_2} - Z_{t_1}, \cdots, Z_{t_n} - Z_{t_{n-1}}$ are independent.

(2) (stationary increments property) The distribution of $Z_{t+s} - Z_t$ is the same for all t.

(3) $Z_0 = 0(P - a.s.)$.

(4) (stochastic continuity)

$$\forall t \geq 0, \forall \epsilon > 0, \lim_{s \to t} P(|Z_s - Z_t| > \epsilon) = 0. \tag{2.1}$$

(5) (cadlag property) There is a subset $\Omega_0 \in F$, $P(\Omega_0) = 1$, such that,

for every $\omega \in \Omega_0$, $Z(t, \omega)$ is right continuous in t and has left limits.

The following are examples of Lévy processes.

Definition 2.2. (Brownian motion). A Lévy process $\{W_t\}$ is called a Brownian motion (or Brownian process, or Wiener process) if the conditional distribution $P(W_{t+h} - W_t | \mathcal{F}_t)$ is $N(0, h)$ for $0 \leq t \leq T$ and $h > 0$.

Remark 2.1. There are many types of characterization of a Brownian motion. For example, the characterization as a Gaussian process, and the characterization as a martingale.

Definition 2.3. (Poisson process). A Lévy process $\{Y_t\}$ is called a Poisson process with intensity parameter λ ($\lambda > 0$) if the conditional distribution $P(Y_{t+h} - Y_t | \mathcal{F}_t)$, $h > 0$, is a Poisson distribution with intensity parameter λh.

The distributions of Lévy processes are characterized in the following form.

Theorem 2.1. *Let $\{Z_t\}$ be a Lévy process, $\mu_t(dz)$ be the distribution of Z_t, and $\phi_t(u)$ be the characteristic function of $\mu_t(dz)$. Then it holds that*

$$\phi_t(u) = (\phi_s(u))^{\frac{t}{s}}, \quad t, s > 0. \tag{2.2}$$

Setting $s = 1$ in this formula, we obtain

$$\phi_t(u) = (\phi(u))^t, \quad t > 0. \tag{2.3}$$

And also, setting $t = 1, s = \frac{1}{n}$, we obtain the following relation:

$$\phi(u) = \phi_1(u) = (\phi_{\frac{1}{n}}(u))^n, \quad n = 1, 2, \ldots. \tag{2.4}$$

This relation is equivalent to the following relation:

$$\mu = \mu_1 = \mu_{\frac{1}{n}} * \mu_{\frac{1}{n}} * \cdots * \mu_{\frac{1}{n}}, \quad n = 1, 2, \ldots. \tag{2.5}$$

This result suggests that the distributions of Lévy processes are related to infinitely divisible distributions.

2.1.2 Infinitely divisible distributions

We give a definition of an infinitely divisible distribution.

Definition 2.4. (Infinitely divisible distribution). A distribution $\mu(dz)$ is said to be infinitely divisible if there exists a distribution $\mu^{(n)}$ for each $n = 1, 2, \ldots$, such that

$$\mu = \mu^{(n)} * \mu^{(n)} * \cdots * \mu^{(n)}, \tag{2.6}$$

where $*$ is a convolution operator.

This definition is equivalent to the following definition.

Definition 2.5. A distribution $\mu(dz)$ is said to be infinitely divisible if the characteristic function $\phi(u)$ of $\mu(dz)$ is expressed, for each $n = 1, 2, \ldots$, in the following form:

$$\phi(u) = (\phi^{(n)}(u))^n, \tag{2.7}$$

where $\phi^{(n)}(u)$ is a characteristic function.

Lévy processes are related to infinitely divisible distributions.

Theorem 2.2. (1) *Let $\{Z_t\}$ be a Lévy process, and let $\mu(dz)$ be the distribution of Z_1. Then the distribution $\mu(dz)$ is infinitely divisible.*
(2) *Suppose that an infinitely divisible distribution $\mu(dz)$ is given. Then there exists a Lévy process such that the distribution of Z_1 is $\mu(dz)$. And such a Lévy process is unique in the sense of distribution.*

By this theorem, distributional properties of a Lévy process are reduced to properties of an infinitely divisible distribution.

Theorem 2.3. (Representation of infinitely divisible distributions).
(1) *Let $\phi(u)$ be a characteristic function of an infinitely divisible distribution. Then $\phi(u)$ is expressed in the following form:*

$$\phi(u) = \exp\left\{ -\frac{\sigma^2}{2}u^2 + \int_{|x|\leq 1} (e^{iux} - 1 - iux)\nu(dx) \right.$$
$$\left. + \int_{|x|>1} (e^{iux} - 1)\nu(dx) + ibu \right\}, \tag{2.8}$$

where $(\sigma^2, \nu(dx), b)$ is a triplet of parameters, which is called a "generating triplet", and satisfies the following [GT] condition:

[GT] *conditions on generating triplet $(\sigma^2, \nu(dx), b)$:*
  (i) $\sigma^2 \geq 0$ *($\sigma = \sqrt{\sigma^2} \geq 0$), $b \in (-\infty, \infty)$.*
  (ii) $\nu(dx)$ *is a measure on $(-\infty, \infty)$, and satisfies the following assumptions:*

$$\nu(\{0\}) = 0 \quad and \quad \int_{|x|>0} (|x|^2 \wedge 1)\nu(dx) < \infty. \tag{2.9}$$

(2) *Suppose that a generating triplet $(\sigma^2, \nu(dx), b)$ which satisfies the [GT] conditions is given. Then the function $\phi(u)$ given by (2.8) is a characteristic function of a distribution $\mu(dz)$, and $\mu(dz)$ is infinitely divisible.*

(3) *The correspondence between an infinitely divisible distribution $\mu(dz)$ and a generating triplet $(\sigma^2, \nu(dx), b)$ is one-to-one. This correspondence is given by* (2.8).

Remark 2.2. (1) The measure $\nu(dx)$ in a triplet $(\sigma^2, \nu(dx), b)$ is called the Lévy measure of an infinitely divisible distribution μ.

(2) From Theorem 2.2 and Theorem 2.3 we know that a Lévy process $\{Z_t\}$ is uniquely determined by the corresponding generating triplet $(\sigma^2, \nu(dx), b)$. Based on this fact, a triplet $(\sigma^2, \nu(dx), b)$ which is corresponding to $\{Z_t\}$ is called the generating triplet of a Lévy process $\{Z_t\}$.

(3) In the representation formula (2.8) for $\phi(u)$, σ^2 and $\nu(dx)$ are uniquely determined. But a constant b and an integrand are not unique.

The integral part of (2.8)

$$\int_{|x|\leq 1} (e^{iux} - 1 - iux)\nu(dx) + \int_{|x|>1} (e^{iux} - 1)\nu(dx)$$

$$= \int_{-\infty}^{\infty} \left(e^{iux} - 1 - iux 1_{\{|x|\leq 1\}}(x) \right) \nu(dx) \tag{2.10}$$

can be expressed in afferent forms. If we select another integrand $c(x)$ such that the integral

$$\int_{-\infty}^{\infty} \left(e^{iux} - 1 - iuxc(x) \right) \nu(dx) \tag{2.11}$$

converges, then $\phi(u)$ of (2.8) is expressed as

$$\phi(u) = \exp\left\{ -\frac{\sigma^2}{2} u^2 + \int_{-\infty}^{\infty} \left(e^{iux} - 1 - iuxc(x) \right) \nu(dx) + ib_c u \right\}, \tag{2.12}$$

where $b_c = b + \int_{-\infty}^{\infty} x \left(c(x) - 1_{\{|x|\leq 1\}}(x) \right) \nu(dx)$. So the representation of a characteristic function $\phi(u)$ depends on a integrand $c(x)$. To make this relation clear, the generating triplet which corresponds to an integrand $c(x)$ is denoted by $(\sigma^2, \nu(dx), b_c)_c$.

In this notation, when we adopt $c(x) = 1_{\{|x|\leq 1\}}(x)$, we omit the subscription c and express $(\sigma^2, \nu(dx), b)$. The convergence of the integral in (2.8) is always certified. In this sense this integrand is a special one, and the expression (2.8) is called "canonical representation".

Example 2.1. If the condition $\int_{|x|>1} |x|\nu(dx) < \infty$ is satisfied, then we can adopt $c(x) \equiv 1$ as an integrand, and the characteristic function is

$$\phi(u) = \exp\left\{ -\frac{\sigma^2}{2} u^2 + \int_{-\infty}^{\infty} \left(e^{iux} - 1 - iux \right) \nu(dx) + ib_1 u \right\}. \tag{2.13}$$

The corresponding generating triplet is $(\sigma^2, \nu(dx), b_1)_1$. This triplet is equivalent to the canonical form $(\sigma^2, \nu(dx), b)$, where

$$b = b_1 - \int_{|x|>1} x\nu(dx). \tag{2.14}$$

It is known that the corresponding infinitely divisible distribution $\mu(dx)$ has a mean and that $b_1 = \int_{-\infty}^{\infty} x\mu(dx)$.

Example 2.2. If the condition $\int_{|x|\leq 1} |x|\nu(dx) < \infty$ is satisfied, then we can adopt $c(x) \equiv 0$ as an integrand, and we obtain

$$\phi(u) = \exp\left\{ -\frac{\sigma^2}{2}u^2 + \int_{-\infty}^{\infty} (e^{iux} - 1)\nu(dx) + ib_0 u \right\}. \tag{2.15}$$

The corresponding generating triplet is $(\sigma^2, \nu(dx), b_0)_0$. This triplet is equivalent to $(\sigma^2, \nu(dx), b)$, where

$$b = b_0 + \int_{|x|\leq 1} x\nu(dx). \tag{2.16}$$

2.1.3   Canonical representation of Lévy processes

As we have seen, a Lévy process corresponds to an infinitely divisible distribution, and this correspondence is one-to-one. So the above theorem suggests that a generating triplet $(\sigma^2, \nu(dx), b)$ determines a Lévy process. In fact the following representation theorem of Lévy processes is known.

[Lévy–Itô decomposition]

A Lévy process Z_t, which corresponds to a generating triplet $(\sigma^2, \nu(dx), b)$ has the following expression:

$$Z_t = \sigma W_t + bt + \int_0^t \int_{|x|>1} x N_p(dudx) + \int_0^t \int_{|x|\leq 1} x\tilde{N}_p(dudx), \tag{2.17}$$

where W_t is a Wiener process, $N_p(dtdx)$ is a Poisson random measure with the compensator $\hat{N}_p(dudx) = du\nu(dx)$, and $\tilde{N}_p(dudx) = N_p(dudx) - \hat{N}_p(dudx)$.

The following are examples of generating triplets and the corresponding Lévy processes.

1) Brownian motion

If the generating triplet is given by

$$(\sigma^2, 0, b), \tag{2.18}$$

then the corresponding Lévy process is

$$Z_t = \sigma W_t + bt. \tag{2.19}$$

This is a Brownian motion with variance σ^2 and drift b. As a special case of $\sigma = 1$, $b = 0$ we obtain the standard Brownian motion (Wiener process).

2) Poisson process

   If the generating triplet is given by

$$(0, \lambda\delta_{\{1\}}(dx), 0)_0, \tag{2.20}$$

the corresponding Lévy process is a Poisson process with intensity λ.

3) Compound Poisson process

   Generalizing the generating triplet of a Poisson process, we consider the following type of triplet:

$$(0, \lambda\rho(dx), 0)_0, \tag{2.21}$$

where λ is a positive constant and $\rho(dx)$ is a probability measure on $(-\infty, \infty)$ with the property $\rho(\{0\}) = 0$. Then the corresponding Lévy process is called a compound Poisson process with intensity λ and jump size measure $\rho(dx)$. We mention here that the above generating triplet is equivalent to a triplet $(0, \lambda\rho(dx), b), b = \int_{|x|\leq 1} \lambda\rho(dx)$.

   A compound Poisson process is a jump type process such that the distribution of the time interval τ of jumps is the exponential distribution with parameter λ (i.e. $P(\tau > t) = \lambda e^{-\lambda t}$) and the distribution of jump size x is $\rho(dx)$.

2.2   Geometric Lévy Process Models

There are many kind of models for stock price processes. In this book we restrict our focus to the geometric Lévy process models. And in this section we see examples of geometric Lévy process models for stock price processes.

   We assume that a probability space $(\Omega, \mathcal{F}, P)$ and a filtration $\{\mathcal{F}_t, 0 \leq t \leq T\}$ are given, and that all processes are defined on this probability space and are $\{\mathcal{F}_t\}$-adapted.

Definition 2.6. A geometric Lévy process (GLP) S_t is

$$S_t = S_0 e^{Z_t}, \tag{2.22}$$

where S_0 is a positive constant (the initial value of the process) and Z_t is a Lévy process with the generating triplet $(\sigma^2, \nu(dx), b)$.

Remark 2.3. A geometric Lévy process is sometimes called an "exponential Lévy process".

We already know some examples of geometric Lévy process models: geometric Brownian motion model, geometric Poisson process model, and geometric compound Poisson process model. Every Lévy process is a candidate for the asset price process model. But from theoretical or practical point of view, several kinds of Lévy processes have been focused on and studied. We list such processes below.

Here are candidates for Lévy process of the GLP model.

(1) Stable process (Mandelbrot (1963) [77]; Fama (1963) [36])
(2) Jump-diffusion process (Merton (1976) [78])
(3) Variance gamma process (Madan–Seneta (1990) [74])
(4) Generalized hyperbolic process (Eberlein–Keller (1995) [32])
(5) CGMY process (Carr–Geman–Madan–Yor (2002) [14])
(6) Normal inverse Gaussian process (Barndorff–Nielsen (1995) [2])
(7) Finite moment log stable process (Carr–Wu (2003) [16])

We next explain some of the geometric Lévy process models which correspond to these Lévy processes.

2.2.1   *The geometric Brownian motion model*

The case where $S_t = S_0 \exp\{Z_t\}$, $Z_t = \sigma W_t + bt$ is the geometric Brownian model. This model is also called the Black–Scholes model. This is the most popular model in use. The generating triplet of Z_t is $(\sigma^2, 0, b)$.

2.2.2   *Geometric compound Poisson models*

Let us consider a case where the price process moves with jumps only. The simplest case is the Poisson process case. Generalizing it, we obtain such models that $S_t = S_0 \exp\{Z_t\}$, where Z_t is a compound Poisson process. The generating triplet of Z_t is $(0, \nu(dx), b)$, where $\nu(dx)$ takes the following form:

$$\nu(dx) = \lambda\rho(dx), \tag{2.23}$$

where λ is a positive constant and $\rho(dx)$ is a probability measure on $(-\infty, \infty)$ such that $\rho(\{0\}) = 0$.

In this case the generating triplet of Z_t has another expression $(0, \lambda\rho(dx), b_0)_0$. Then the characteristic function $\phi(u)$ of Z_t is expressed in the following form:

$$\phi(u) = E[e^{iuZ_1}] = \exp(ib_0u + \lambda(\hat{\rho}(u) - 1)), \qquad (2.24)$$

where

$$\hat{\rho}(u) = \int_{-\infty}^{\infty} e^{iux}\rho(dx). \qquad (2.25)$$

From this expression we infer that if S_t has k-th moment then it holds that

$$E[(S_t)^k] = (S_0)^k E[e^{kZ_t}] = (S_0)^k \exp\left(t\left(b_0k + \lambda(\int_{-\infty}^{\infty} e^{kx}\rho(dx) - 1)\right)\right). \qquad (2.26)$$

Let $M_\rho(\theta)$ be the moment-generating function of ρ:

$$M_\rho(\theta) = \int_{-\infty}^{\infty} e^{\theta x}\rho(dx). \qquad (2.27)$$

Then we know that S_t has the k-th moment for k such that $M_\rho(k)$ exists.

1) Discrete type Lévy measure case

A simple one is the case where $\nu(dx) = \lambda\rho(dx)$ is of the following form:

$$\nu(dx) = \lambda\rho(dx) = \lambda\sum_{j=1}^{d} p_j\delta_{a_j}(dx), \quad p_j > 0, j = 1, 2, \ldots, d, \quad \sum_{j=1}^{d} p_j = 1. \qquad (2.28)$$

In this case the characteristic function $\phi(u)$ of Z_t is

$$\phi(u) = \exp\left(ib_0u + \lambda\sum_{j=1}^{d} p_j(e^{iua_j} - 1)\right), \qquad (2.29)$$

and the moment-generating function of ρ is

$$M_\rho(\theta) = \sum_{j=1}^{d} p_j e^{a_j\theta}. \qquad (2.30)$$

Since $M_\rho(\theta)$ is well defined for $\theta \in (-\infty, \infty)$, S_t has moments of all degrees.

2) Normal distribution type Lévy measure case

We see the case where $\nu(dx)$ is given by

$$\nu(dx) = \lambda g(x; m, v)dx = \lambda\frac{1}{\sqrt{2\pi v}}\exp\left(-\frac{(x - m)^2}{2v}\right)dx, \qquad (2.31)$$

where λ, m, v are constants such that $\lambda > 0$, $m \in (-\infty, \infty)$, $v > 0$. Then the generating triplet is expressed in the form of $(0, \nu(dx), b_0)_0$ and the characteristic function $\phi(u)$ of Z_t is

$$\phi(u) = \exp\left(ib_0 u + \lambda\left(\hat{g}(u) - 1\right)\right), \tag{2.32}$$

where

$$\hat{g}(u) = \int_{-\infty}^{\infty} e^{iux} g(x; m, v) dx = \exp\left(imu - \frac{1}{2}vu^2\right). \tag{2.33}$$

The moment-generating function of ρ is

$$M_\rho(\theta) = \int_{-\infty}^{\infty} e^{\theta x} g(x; m, v) dx = \exp\left(m\theta + \frac{1}{2}v\theta^2\right), \tag{2.34}$$

and S_t has moments of all degrees.

2.2.3 *Jump-diffusion models*

A Lévy process $\{Z_t\}$ is called a jump-diffusion process if its generating triplet is expressed in the form of $(\sigma^2, \nu(dx), b_0)_0$ and $\nu(dx) = c\rho(dx)$. In this case the model $S_t = S_0 \exp\{Z_t\}$ is called a geometric jump-diffusion model. A jump-diffusion type Lévy process $\{Z_t\}$ is expressed in the following form:

$$Z_t = \sigma W_t + b_0 t + J_t, \tag{2.35}$$

where J_t is a compound Poisson process with the Lévy measure $\nu(dx) = c\rho(dx)$.

2.2.4 *Geometric variance gamma models*

Suppose that $\{Z_t\}$ is a variance gamma process, namely that the generating triplet of $\{Z_t\}$ is in the form of $(0, \nu(dx), b_0)_0$ and

$$\nu(dx) = C \frac{\left(I_{\{x<0\}}(x)e^{-c_1|x|} + I_{\{x>0\}}(x)e^{-c_2|x|}\right)}{|x|} dx, \tag{2.36}$$

where C, c_1, and c_2 are positive constants. Then $S_t = S_0 \exp\{Z_t\}$ is called a geometric variance gamma model. A variance gamma process is determined by four parameters (C, c_1, c_2, b_0), so we use a notation $VG(C, c_1, c_2, b_0)$ for a variance gamma process.

The characteristic function is

$$\phi(u) = E[e^{iuZ_1}]$$

$$= \exp\left\{ib_0 u - C\left(\log(1 + \frac{iu}{c_1}) + \log(1 - \frac{iu}{c_2})\right)\right\}$$

$$= e^{ib_0 u}\left(\frac{1}{\left(1 + \frac{iu}{c_1}\right)\left(1 - \frac{iu}{c_2}\right)}\right)^C. \tag{2.37}$$

2.2.5   Geometric stable process models

Suppose that the Lévy measure ν is given in the following form:

$$\nu(dx) = \begin{cases} c_1 \dfrac{1}{|x|^{1+\alpha}} dx, & x < 0, \\[2ex] c_2 \dfrac{1}{x^{1+\alpha}} dx, & x > 0, \end{cases} \tag{2.38}$$

where α, c_1, c_2 are constants such that $0 < \alpha < 2$, $c_1 \geq 0$, $c_2 \geq 0$ and $c_1 + c_2 > 0$. Then the Lévy process with the generating triplet $(0, \nu(dx), b)$, $-\infty < b < \infty$, is called an α-stable process. It is known that α-stable processes has the k-th moment for $k < \alpha$ but not for $k \geq \alpha$.

To see the characteristic function of α-stable process, we introduce new parameters $(\alpha, \tilde{c}, \beta, \tau)$ while $0 < \alpha < 2$, $\tilde{c} > 0$, $-1 \leq \beta \leq 1$, $-\infty < \tau < \infty$, which are defined by the following formulas:

$$\tilde{c} = c_1 + c_2, \tag{2.39}$$

$$\beta = \frac{c_2 - c_1}{c_1 + c_2}, \tag{2.40}$$

$$\tau = \begin{cases} b_0 = b - \int_{|x|\leq 1} x\nu(dx) = b - \frac{c_2 - c_1}{1-\alpha}, & if \quad 0 < \alpha < 1, \\ b + c_0\tilde{c}, & if \quad \alpha = 1, \\ b_1 = b - \int_{|x|>1} x\nu(dx) = b - \frac{c_2 - c_1}{1-\alpha}, & if \quad 1 < \alpha < 2, \end{cases} \tag{2.41}$$

where $c_0 = \int_0^1 r^{-2}(\sin r - r)dr + \int_1^\infty r^{-2}\sin r\, dr$. Using these new parameters, we obtain the following characteristic function:

$$\phi(u) = \phi_{stable}(u) = \exp(\psi(u)),$$

$$\psi(u) = \begin{cases} \Gamma(-\alpha)(\cos\frac{\pi\alpha}{2})\tilde{c}|u|^\alpha\left(1 - i\beta\tan\frac{\pi\alpha}{2}\mathrm{sgn}(u)\right) + i\tau u & for \quad \alpha \neq 1 \\ -\frac{\pi}{2}\tilde{c}|u|\left(1 + i\beta\frac{2}{\pi}\mathrm{sgn}(u)\log|u|\right) + i\tau u & for \quad \alpha = 1. \end{cases} \tag{2.42}$$

Remark 2.4. The inverse transform of the above parameters is given by

$$c_1 = \frac{\tilde{c} - \tilde{c}\beta}{2}, \tag{2.43}$$

$$c_2 = \frac{\tilde{c} + \tilde{c}\beta}{2}, \tag{2.44}$$

$$b = \begin{cases} \tau + \frac{c_2 - c_1}{1 - \alpha} = \tau + \frac{\tilde{c}\beta}{1 - \alpha}, & for \quad \alpha \neq 1, \\ \tau - c_0\tilde{c}, & for \quad \alpha = 1. \end{cases} \tag{2.45}$$

The characteristic function has another expression. Introduce a new parameter $\sigma > 0$ by

$$\begin{cases} \sigma = \left(-\Gamma(-\alpha)(\cos\frac{\pi\alpha}{2})\tilde{c} \right)^{\frac{1}{\alpha}} & for \quad \alpha \neq 1, \\ \sigma = \frac{\pi}{2}\tilde{c} & for \quad \alpha = 1. \end{cases} \tag{2.46}$$

Then $\psi(u)$ is expressed in the following form:

$$\psi(u) = \begin{cases} -\sigma^\alpha |u|^\alpha \left(1 - i\beta \tan\frac{\pi\alpha}{2}\mathrm{sgn}(u) \right) + i\tau u & for \quad \alpha \neq 1 \\ -\sigma|u| \left(1 + i\beta\frac{2}{\pi}\mathrm{sgn}(u)\log|u| \right) + i\tau u & for \quad \alpha = 1 \end{cases}. \tag{2.47}$$

The stable process determined by parameters $(\alpha, \sigma, \beta, \tau)$ is traditionally denoted by $S_\alpha(\sigma, \beta, \tau)$.

2.2.6 *Geometric CGMY models*

The Lévy measure of a CGMY process is

$$\nu(dx) = C \left(\frac{I_{\{x<0\}}(x)\exp(-G|x|) + I_{\{x>0\}}(x)\exp(-M|x|)}{|x|^{(1+Y)}} \right) dx, \tag{2.48}$$

where $C > 0, G \geq 0, M \geq 0, Y < 2$, and it is assumed that $G > 0$ and $M > 0$ if $Y \leq 0$. The generating triplet of this process is $(0, \nu(dx), b)$. We remark here that if $Y = 0$ this model is a variance gamma model, and that if $G = M = 0$ and $0 < Y < 2$, then this is a symmetric stable process model.

We assume that $G > 0$ and $M > 0$. Then the generating triplet of a CGMY process has an expression of the form $(0, \nu(dx), b_1)_1$. The characteristic function of a CGMY process is in the following form:

(1) The case of $Y = 0$ (variance gamma process): In this case the generating triplet has an expression $(0, \nu(dx), b_0)_0$, and the characteristic function is

$$\phi_{VG}(u) = e^{ib_0 u} \left(\frac{1}{\left(1 + \frac{iu}{G}\right)\left(1 - \frac{iu}{M}\right)} \right)^C, \tag{2.49}$$

where b_0 is given by

$$b_0 = b_1 - \int_{-\infty}^{\infty} x\nu(dx) = b_1 - C\left(\frac{1}{M} - \frac{1}{G}\right). \qquad (2.50)$$

(2) The case of $Y < 1$ ($Y \neq 0$): In this case the characteristic function is

$$\phi_{CGMY}(u) = \exp\left\{ib_0 u + C\Gamma(-Y)\left((M-iu)^Y - M^Y \right.\right.$$
$$\left.\left. +(G+iu)^Y - G^Y\right)\right\}, (2.51)$$

where $\Gamma(\cdot)$ is a gamma function and b_0 is given by

$$b_0 = b_1 + C\Gamma(-Y)Y(M^{Y-1} - G^{Y-1}). \qquad (2.52)$$

(3) The case of $Y = 1$: In this case the characteristic function is

$$\phi_{CGMY}(u) = \exp\left\{ib_1 u + C\left((M-iu)\log(1 - \frac{iu}{M})\right.\right.$$
$$\left.\left. +(G+iu)\log(1 + \frac{iu}{G})\right)\right\}. \qquad (2.53)$$

(4) The case of $1 < Y < 2$:

$$\phi_{CGMY}(u) = \exp\left\{i\left(b_1 + C\Gamma(-Y)Y\left(M^{Y-1} - G^{Y-1}\right)\right)u\right.$$
$$\left. +C\Gamma(-Y)\left((M-iu)^Y - M^Y + (G+iu)^Y - G^Y\right)\right\}.$$
$$(2.54)$$

Remark 2.5. From the above results, we can see that the characteristic function $\phi(u)$ has the following form:

$$\phi(u) = \exp\left\{i\left(b_1 + C\Gamma(-Y)Y\left(M^{Y-1} - G^{Y-1}\right)\right)u\right.$$
$$\left. +C\Gamma(-Y)\left((M-iu)^Y - M^Y + (G+iu)^Y - G^Y\right)\right\}. \quad (2.55)$$

In this formula we suppose that, at the singular point of $\Gamma(Y)$, the function $\phi(u)$ is defined so that $\phi(u)$ is continuous in Y.

2.3 Doléans-Dade Exponential

A geometric Lévy process S_t can also have the following expression:

$$S_t = S_0 \mathcal{E}(\hat{Z})_t, \qquad (2.56)$$

where $\mathcal{E}(\hat{Z})_t$ is the Doléans-Dade exponential of $\hat{Z}_t$. And $\hat{Z}_t$ is defined by

$$\hat{Z}_t = Z_t + \frac{1}{2}\langle Z^c\rangle_t + \sum_{s\in(0,t]}\{e^{\triangle Z_s} - 1 - \triangle Z_s\} \tag{2.57}$$

$$= Z_t + \frac{1}{2}\sigma^2 t + \int_0^t\int_{-\infty}^{\infty}(e^x - 1 - x)N_p(dudx) \tag{2.58}$$

$$= \sigma W_t + \tilde{b}t + \int_0^t\int_{|x|\leq 1}(e^x - 1)\tilde{N}_p(dudx)$$

$$+ \int_0^t\int_{|x|>1}(e^x - 1)N_p(dudx) \tag{2.59}$$

$$= \sigma W_t + \tilde{b}'t + \int_0^t\int_{-\infty}^{\infty}(e^x - 1)\tilde{N}_p(dudx), \tag{2.60}$$

where

$$\tilde{b} = b + \frac{1}{2}\sigma^2 + \int_{|x|\leq 1}((e^x - 1 - x))\,\nu(dx), \tag{2.61}$$

and

$$\tilde{b}' = b + \frac{1}{2}\sigma^2 + \int_{-\infty}^{\infty}((e^x - 1 - x1_{\{|x|\leq 1\}}(x)))\,\nu(dx). \tag{2.62}$$

The generating triplet of $\hat{Z}_t$, $(\hat{\sigma}^2, \hat{\nu}(dx), \hat{b})$, is given by

$$\hat{\sigma}^2 = \sigma^2 \tag{2.63}$$

$$\hat{\nu}(dx) = (\nu \circ J^{-1})(dx), \quad J(x) = e^x - 1, \tag{2.64}$$

$$(\quad i.e. \quad \hat{\nu}(A) = \int_{-\infty}^{\infty}1_A(x)(e^x - 1)\nu(dx) \quad)$$

$$\hat{b} = b + \frac{1}{2}\sigma^2 + \int_{-\infty}^{\infty}((e^x - 1)1_{\{e^x - 1\leq 1\}} - x1_{\{|x|\leq 1\}}(x))\,\nu(dx). \tag{2.65}$$

Remark 2.6. (i) S_t satisfies the following equation:

$$dS_t = S_{t-}d\hat{Z}_t. \tag{2.66}$$

(ii) It holds that

$$supp \; \hat{\nu} \subset (-1, \infty). \tag{2.67}$$

(iii) If $\nu(dx)$ has a density $n(x)$, then $\hat{\nu}(dx)$ has a density $\hat{n}(x)$, and $\hat{n}(x)$ is given by

$$\hat{n}(x) = \frac{1}{1 + x}n(\log(1 + x)). \tag{2.68}$$

(iv) Relations between Z_t and $\hat{Z}_t$ and their economic meanings are discussed more precisely in Chapter 4.

Notes

There are many books available for the study of Lévy processes. We list some of them below:

Applebaum, D. (2004) [1],
Barndorff-Nielsen, O.E., Mikosch, T., and Resnick, S.I. (Ed.) (2001) [4],
Bertoin, J. (1996) [8],
Sato, K. (1999) [109].

Among these books, I have mainly referred to Sato, K. (1999) [109]. If a reader of this book finds terminology that is not precisely described, please refer to Sato, K. (1999) [109] for details.

The following books are recommended for the study of Lévy processes with relation to mathematical finance:

Cont, R. and Tankov, P. (2004a) [25],
Schoutens, W. (2003) [112].

Chapter 3

Equivalent Martingale Measures

The concept of equivalent martingale measure is very important in mathematical finance theory.

3.1 Equivalent Martingale Measure Methods

The equivalent martingale measure method is one of the most powerful methods for option pricing. If a market satisfies the no-arbitrage assumption, then an equivalent martingale measure exists. And conversely, if an equivalent martingale measure exists, then the market is arbitrage-free. (The First Fundamental Theorem in Mathematical Finance.) Under the assumption that a market is arbitrage-free, if the market is complete, then the equivalent martingale measure is unique. Conversely, if a market has a unique equivalent martingale measure, then the market is arbitrage-free and complete. (Second Fundamental Theorem in Mathematical Finance.) On the other hand, if a market is arbitrage-free and incomplete, then the market has an infinite number of equivalent martingale measures. Conversely, if a market has more than one equivalent martingale measure, then the market is arbitrage-free and incomplete.

In the case that a market is arbitrage-free and incomplete, there are many equivalent martingale measures. So we have to select the most suitable martingale measure in order to apply the martingale measure method to incomplete markets.

The geometric Lévy process model is now very popular in mathematical finance theory. This model is an incomplete market model, and it has many equivalent martingale measures. For the construction of an option pricing model, we have to select a suitable martingale measure. Once an equivalent martingale measure Q is selected, then the price $\pi(X)$ of an option X is given by

$$\pi(X) = E_Q[e^{-rT}X], \qquad (3.1)$$

where r is the interest rate of a risk-free asset and T is a maturity. This is the idea of the equivalent martingale measure method.

3.2   Equivalent Martingale Measures for Geometric Lévy Processes

Suppose that a probability space (Ω, F, P) and a filtration $\{\mathcal{F}_t, t \geq 0\}$ are given as usual. Let S_t be an underlying asset process, which is defined on the space (Ω, F, P) and $\{\mathcal{F}_t\}$-adaptive, and let $r(> 0)$ be the interest rate of a risk-free asset.

Definition 3.1. A probability measure Q defined on (Ω, F) is called an equivalent martingale measure of S_t if $Q \sim P$ (equivalent) and $e^{-rt}S_t$ is $\{\mathcal{F}_t\}$-martingale.

Remark 3.1. The original probability P is called "physical probability".

3.2.1   *Candidates for suitable equivalent martingale measure*

Several kinds of candidate for a suitable equivalent martingale measure are proposed. We list some of them below:
(1) Minimal Martingale Measure (MMM) (Föllmer–Schweizer (1991) [39]),
(2) Variance Optimal Martingale Measure (VOMM) (Schweizer (1995) [113]),
(3) Mean Correcting Martingale Measure (MCMM),
(4) Esscher Martingale Measure (ESSMM) (Gerber–Shiu (1994) [46]; Bühlmann–Delbaen–Embrechts–Shiryaev (1996) [12]),
(5) Minimal Entropy Martingale Measure (MEMM) (Miyahara (1996a) [81]; Frittelli (2000a) [40]),
(6) Utility-Based Martingale Measure (U-MM).

For the selection of the suitable martingale measure, we have to study the appropriateness of the models. This should be done from two different view points. One is the theoretical point of view, and the other is the empirical point of view. The study from the theoretical point of view is done in Chapters 4, 5, 6, and 7. The calibration problem, which is studied in Chapters 8 and 9, is connecting the theory to applications and empirical analysis.

3.3 Methods for Construction of Martingale Measures

Suppose that an underlying asset process S_t is given. Let us consider the construction problem of equivalent martingale measures for S_t.

Two different methods are known. One is the Esscher transformation method, and the other one is the minimal distance method. The Esscher transformation method is well known in risk theory.

The Esscher transformation is a very useful technique to obtain a reasonable equivalent martingale measure, and it is related to the corresponding risk process. (See [35], [46], [12], and [62].) We shall see details of this method in Chapter 4.

The second method is the minimal distance method. This method is related to the maximization of expected utility. We will see details of this method in Chapters 5 and 6.

The variance optimal martingale measure (VOMM) and the minimal entropy martingale measure (MEMM) are important examples of the minimal distance martingale measures. Here we explain these two martingale measures for the most simple, discrete model.

3.3.1 *Variance optimal martingale measure (VOMM)*

Let $(\Omega, \mathcal{F}, P)$ be a finite probability space, and set $\Omega = \{\omega_1, \dots, \omega_n\}$ and $P(\{\omega_j\}) = p_j, j = 1, \dots, n$. We consider a one-period model defined on this probability space. Let $S_t(\omega), t = 0, 1$, be the underlying price process of this model.

The VOMM $\tilde{Q}$ is an equivalent martingale measure for S_t which satisfies the following minimization condition:

$$\int_\Omega \left(\frac{d\tilde{Q}}{dP} \right)^2 dP = \min_{\{Q:EMM\}} \{ \int_\Omega \left(\frac{dQ}{dP} \right)^2 dP \}. \tag{3.2}$$

Setting $Q(\{\omega_j\}) = q_j, j = 1, \dots, n$, we obtain the following problem:

$$Minimize \quad f(q_1, \dots, q_n) = \sum_{j=1}^{n} \left(\frac{q_j}{p_j} \right)^2 \tag{3.3}$$

subject to

$$\sum_{j=1}^{n} \triangle S_1(\omega_j) q_j = 0, \quad \triangle S_1(\omega_j) = S_1(\omega_j) - S_0, j = 1, \ldots, n, \qquad (3.4)$$

and

$$\sum_{j=1}^{n} q_j = 1. \qquad (3.5)$$

Introducing Lagrange multipliers λ and μ, we obtain the following Lagrange function:

$$L(q_1, \ldots, q_n; \lambda, \mu) = f(q_1, \ldots, q_n) - \lambda \sum_{j=1}^{n} \triangle S_1(\omega_j) q_j - \mu (\sum_{j=1}^{n} q_j - 1). \quad (3.6)$$

The Lagrange equations are

$$\frac{2q_j}{p_j^2} - \lambda \triangle S_1(\omega_j) - \mu = 0, \quad j = 1, 2, \ldots, n. \qquad (3.7)$$

From this we obtain

$$q_j = \frac{p_j^2}{2} (\lambda \triangle S_1(\omega_j) + \mu), j = 1, 2, \ldots, n, \qquad (3.8)$$

and from this and (3.4), we have the following equation:

$$\sum_{j=1}^{n} \triangle S_1(\omega_j) \frac{p_j^2}{2} (\lambda \triangle S_1(\omega_j) + \mu) = 0. \qquad (3.9)$$

We also obtain the following equation from (3.8) and (3.5):

$$\sum_{j=1}^{n} \frac{p_j^2}{2} (\lambda \triangle S_1(\omega_j) + \mu) = 1. \qquad (3.10)$$

Solving equations (3.9) and (3.10) for λ and μ, we obtain

$$\lambda = \frac{-2 \left(\sum_{j=1}^{n} p_j^2 \triangle S_1(\omega_j) \right)}{\left(\sum_{j=1}^{n} p_j^2 \right) \left(\sum_{j=1}^{n} p_j^2 \triangle S_1(\omega_j)^2 \right) - \left(\sum_{j=1}^{n} p_j^2 \triangle S_1(\omega_j) \right)^2} \quad (3.11)$$

$$\mu = \frac{2 \left(\sum_{j=1}^{n} p_j^2 \triangle S_1(\omega_j)^2 \right)}{\left(\sum_{j=1}^{n} p_j^2 \right) \left(\sum_{j=1}^{n} p_j^2 \triangle S_1(\omega_j)^2 \right) - \left(\sum_{j=1}^{n} p_j^2 \triangle S_1(\omega_j) \right)^2}. \quad (3.12)$$

And from (3.8) the VOMM is

$$\tilde{q}_k = \frac{\left[\left(\sum_{j=1}^{n} p_j^2 \triangle S_1(\omega_j) \right) \triangle S_1(k) - \left(\sum_{j=1}^{n} p_j^2 \triangle S_1(\omega_j)^2 \right) \right] p_k^2}{\left(\sum_{j=1}^{n} p_j^2 \right) \left(\sum_{j=1}^{n} p_j^2 \triangle S_1(\omega_j)^2 \right) - \left(\sum_{j=1}^{n} p_j^2 \triangle S_1(\omega_j) \right)^2}, \quad (3.13)$$

$$k = 1, 2, \ldots, n. \qquad (3.14)$$

Remark 3.2. $\tilde{q}_k$ of (3.14) may be negative. Sometimes therefore, the VOMM is a signed measure.

3.3.2 *Minimal entropy martingale measure (MEMM)*

We first give the definition of relative entropy.

Definition 3.2. Let P and Q be two probability measures given on a same probability space. Then $H(Q|P)$ defined by

$$H(Q|P) = \begin{cases} \int_\Omega \log[\frac{dQ}{dP}]dQ, & if \quad Q \ll P, \\ \infty, & otherwise, \end{cases} \tag{3.15}$$

is called the relative entropy of Q with respect to P.

We can now give the definition of the minimal entropy martingale measure (MEMM) as following:

Definition 3.3. (Minimal Entropy Martingale Measure (MEMM)).
If an equivalent martingale measure P^* satisfies

$$H(P^*|P) \leq H(Q|P) \quad \forall Q : \text{ equivalent martingale measure}, \tag{3.16}$$

then P^* is called the minimal entropy martingale measure (MEMM) *of S_t.*

Next we consider the MEMM under the same situation with the VOMM. The problem to be solved is:

$$minimize \quad f(q_1,\ldots,q_n) = \sum_{j=1}^n q_j \log \frac{q_j}{p_j} \tag{3.17}$$

subject to

$$\sum_{j=1}^n \triangle S_1(\omega_j)q_j = 0, \tag{3.18}$$

and

$$\sum_{j=1}^n q_j = 1. \tag{3.19}$$

Similarly in the case of VOMM, we obtain the following Lagrange equation:

$$\log\left(\frac{q_j}{p_j}\right) + 1 - \lambda\triangle S_1(\omega_j) - \mu = 0, \quad j = 1,\ldots,n, \tag{3.20}$$

and from this formula we obtain:

$$q_j = p_j e^{\lambda\triangle S_1(\omega_j)+\mu-1}, j = 1, 2, \ldots, n. \tag{3.21}$$

So from the martingale condition (3.18) we get:

$$\sum_{j=1}^{n} p_j \triangle S_1(\omega_j) e^{\lambda \triangle S_1(\omega_j)} = 0. \tag{3.22}$$

From this formula λ is determined. And from (3.19), (3.21), and (3.22), we obtain:

$$e^{\mu-1} = \frac{1}{\sum_{j=1}^{n} p_j e^{\lambda \triangle S_1(\omega_j)}} = \frac{1}{E_P[e^{\lambda \triangle S_1}]}. \tag{3.23}$$

And finally from (3.21) and (3.23), we obtain the solution:

$$q_j^* = \frac{p_j e^{\lambda \triangle S_1(\omega_j)}}{E_P[e^{\lambda \triangle S_1}]}, j = 1, 2, \ldots, n, \tag{3.24}$$

where λ is determined by (3.22).

Remark 3.3. From the above result (3.24) we know that the MEMM is an Esscher-transformed martingale measure. We shall study the details of Esscher-transformed martingale measures in Chapter 4.

Notes

The minimal martingale measure (MMM), which was introduced by Föllmer and Schweizer (1991) [39], was the first candidate for a suitable equivalent martingale measure. After that several candidates have been offered, including the Esscher martingale measure (ESSMM) by Gerber and Shiu (1994) [46], the variance optimal martingale measure (VOMM) by Schweizer (1996) [114]), and the minimal entropy martingale measure (MEMM(=CMM)) by Miyahara (1996a) [81]. The MEMM was first named "canonical martingale measure (CMM)" in [81]. The name MEMM was introduced by Frittelli (2000a) [40].

The MMM and VOMM are both related to L^2-hedging theory. The ESSMM is related to the Esscher transformation, which is popular in risk theory. The MEMM was first introduced in relation with large deviation theory (see [81]) and later it was shown that the MEMM is related to exponential utility function (see [40]).

The MEMM for the geometric Lévy process is obtained by the Esscher transformation using the simple-return process (see [44], [89], [90], and

[53]). In this sense the ESSMM and the MEMM are in the same group. (See Chapter 4.)

On the other hand, the VOMM and the MEMM are in another group of martingale measures in the sense that both of them are in the class of minimal distance martingale measures. The VOMM is the minimal (signed) martingale measure in the sense of L^2, and the MEMM is minimal in the sense of relative entropy. (See Chapter 6.)

Since the MEMM is contained in both groups of martingale measures, we can suppose that the MEMM is a special martingale measure. In fact, we shall see many useful properties of the MEMM in Chapters 7, 8, and 9.

Chapter 4

Esscher-Transformed Martingale Measures

4.1 Esscher Transformation

The Esscher transformation is very popular in the fields of risk management and the actuarial profession. Esscher introduced the idea of risk function and transformed risk function for the calculation of collective risk in [35]. His idea has been developed by many researchers (for example, [46], [12], and [62]), and applied to option pricing theory.

Below we give the definitions of Esscher transformation and of the Esscher-transformed martingale measure.

Definition 4.1. Let R be a risk variable, namely a random variable defined on a probability space $(\Omega, \mathcal{F}, P)$, and h be a constant. Then the probability measure $P_{R,h}^{(ESS)}$ defined by

$$\frac{dP_{R,h}^{(ESS)}}{dP}|_{\mathcal{F}} = \frac{e^{hR}}{E[e^{hR}]} \tag{4.1}$$

is called the Esscher-transformed measure of P by a random variable R and a constant h, and this transformation of measures is called the Esscher transformation by a random variable R and a constant h.

Definition 4.2. Let $R_t, 0 \le t \le T$, be a risk process, namely a stochastic process. Then the probability measure $P_{R_{[0,T]},h}^{(ESS)}$ defined by

$$\frac{dP_{R_{[0,T]},h}^{(ESS)}}{dP}|_{\mathcal{F}} = \frac{e^{hR_T}}{E[e^{hR_T}]} \tag{4.2}$$

is called the Esscher-transformed measure of P by a process R_t and a constant h, and this transformation is called the Esscher transformation by a process R_t and a constant h. (We remark that $P_{R_{[0,T]},h}^{(ESS)} = P_{R_T,h}^{(ESS)}$.)

29

Definition 4.3. In Definition 4.2, if the constant h is chosen so that the $P_{R_{[0,T]},h}^{(ESS)}$ is a martingale measure for an underlying price process S_t, then $P_{R_{[0,T]},h}^{(ESS)}$ is called the "Esscher-transformed martingale measure" (ESSTMM) of S_t by a risk process R_t, and it is denoted by $P_{R_{[0,T]}}^{(ESSMM)}$ or $P_{R_T}^{(ESSMM)}$.

4.2 Esscher-Transformed Martingale Measure for Geometric Lévy Processes

Suppose that a probability space $(\Omega, \mathcal{F}, P)$ and a filtration $\{\mathcal{F}_t, 0 \leq t \leq T\}$ are given, and that a price process $S_t = S_0 e^{Z_t}$ of a stock is defined on this probability space, where Z_t is a Lévy process.

As usual we assume that $\mathcal{F}_t = \sigma(S_s, 0 \leq s \leq t) = \sigma(Z_s, 0 \leq s \leq t)$ and $\mathcal{F} = \mathcal{F}_T$. A probability measure Q on $(\Omega, \mathcal{F})$ is called an equivalent martingale measure of S_t if $Q \sim P$ and $e^{-rt} S_t$ is $(\mathcal{F}_t, Q)$-martingale, where r is a constant interest rate.

4.2.1 *Simple-return process and compound-return process*

When we are given a certain risk process R_t, we obtain a corresponding Esscher-transformed martingale measure if it exists.

As we have seen in Chapter 2, the GLP has two kinds of representation such that

$$S_t = S_0 e^{Z_t} = S_0 \mathcal{E}(\hat{Z})_t. \tag{4.3}$$

The processes Z_t and $\hat{Z}_t$ are candidates for a risk process. Z_t is called the "compound-return process" of S_t and $\hat{Z}_t$ is called the "simple-return process" of S_t.

Corresponding to these two risk processes, we obtain two kinds of Esscher-transformed martingale measures. The first one is the "compound-return Esscher-transformed martingale measure". This martingale measure was first introduced by Gerber and Shiu (1994) [46], and this measure is called the "Esscher martingale measure" (ESSMM). The second one is the "simple-return Esscher-transformed martingale measure". It is known that this martingale measure is identified with the "minimal entropy martingale measure" (MEMM). We shall see this fact in Chapter 6. The MEMM has been discussed in [81], [40], and [44].

To investigate the economic meaning of those risk processes, we review the discrete time approximation of a geometric Lévy process.

Set the following:

$$S_k^{(n)} = S_{k/2^n}, \quad k = 1, 2, \dots. \tag{4.4}$$

According to the two kinds of expression (4.3) of S_t, we obtain two kinds of approximation formula.

The first is:

$$S_k^{(n)} = S_0 e^{Z_k^{(n)}}, \quad k = 1, 2, \dots, \tag{4.5}$$

where $Z_k^{(n)} = Z_{k/2^n}$.

The second approximation is:

$$S_k^{(n)} = S_0 \mathcal{E}(Y^{(n)})_k, \quad k = 1, 2, \dots, \tag{4.6}$$

where $\mathcal{E}(Y^{(n)})_k$ is the discrete time Doléans-Dade exponential of $Y_k^{(n)}$,

$$\mathcal{E}(Y^{(n)})_k = \prod_{j=1}^{k} \left(1 + (Y_j^{(n)} - Y_{j-1}^{(n)})\right) \tag{4.7}$$

and $Y_k^{(n)}$ is defined from the following relation:

$$e^{Z_k^{(n)}} = \mathcal{E}(Y^{(n)})_k = \prod_{j=1}^{k} \left(1 + (Y_j^{(n)} - Y_{j-1}^{(n)})\right), \quad k = 1, 2, \dots. \tag{4.8}$$

So we obtain

$$e^{\triangle Z_k^{(n)}} = e^{Z_k^{(n)} - Z_{k-1}^{(n)}} = \left(1 + (Y_k^{(n)} - Y_{k-1}^{(n)})\right) = 1 + \triangle Y_k^{(n)}. \tag{4.9}$$

From this we obtain

$$\frac{\triangle S_k^{(n)}}{S_{k-1}^{(n)}} = \frac{S_k^{(n)} - S_{k-1}^{(n)}}{S_{k-1}^{(n)}} = \frac{S_k^{(n)}}{S_{k-1}^{(n)}} - 1 = e^{\triangle Z_k^{(n)}} - 1 = \triangle Y_k^{(n)}, \tag{4.10}$$

and we know that $\triangle Y_k^{(n)}$ is the simple-return process of $S_k^{(n)}$.

On the other hand, we obtain the following formula for $\triangle Z_k^{(n)}$:

$$\triangle Z_k^{(n)} = \log\left(1 + \triangle Y_k^{(n)}\right) = \log\left(1 + \frac{\triangle S_k^{(n)}}{S_{k-1}^{(n)}}\right), \tag{4.11}$$

and we know that $\triangle Z_k^{(n)}$ is the compound-return process of $S_k^{(n)}$.

For $t \in (\frac{k-1}{2^n}, \frac{k}{2^n})$ we define

$$Z_t^{(n)} = Z_k^{(n)}, \quad Y_t^{(n)} = Y_k^{(n)}. \tag{4.12}$$

It is easy to see that the process $Z_t^{(n)}$ converges to Z_t when n goes to ∞. On the other hand we can see that the process $Y_t^{(n)}$ converges to $\hat{Z}_t$.

As we have already seen, S_t satisfies the following stochastic differential equation:

$$dS_t = S_{t-}d\hat{Z}_t. \tag{4.13}$$

From this it follows that:

$$d\hat{Z}_t = \frac{dS_t}{S_{t-}}. \tag{4.14}$$

Comparing the formulas (4.10) and (4.14), we know that the process $Y_t^{(n)}$ is an approximation process of $\hat{Z}_t$, and that $Y_t^{(n)}$ converges to $\hat{Z}_t$.

Based on the observation above, it is natural for us to give the following definition.

Definition 4.4. The process $\hat{Z}_t$ is called the "simple-return process" of S_t, and the process Z_t is called the "compound-return process" of S_t.

Remark 4.1. The terms "simple-return" and "compound-return" were introduced in [12] p.294.

4.2.2 *Two kinds of Esscher-transformed martingale measure*

Suppose that Z_t is adopted as a risk process. In this case, if the corresponding Esscher-transformed martingale measure $P_{Z_{[0,T]}}^{(ESS)}$ is well defined, then this martingale measure should be called the "compound-return Esscher-transformed martingale measure". This is the Esscher martingale measure introduced by Gerber and Shiu in [46], and the term "Esscher martingale measure" usually means this compound-return Esscher-transformed martingale measure $P_{Z_{[0,T]}}^{(ESS)}$.

Next we consider the case where $\hat{Z}_t$ is adopted as a risk process. If the corresponding Esscher-transformed martingale measure $P_{\hat{Z}_{[0,T]}}^{(ESS)}$ exists, then this martingale measure should be called the "simple-return Esscher-transformed martingale measure".

In [62] the following results have been obtained.

Proposition 4.1. ([62], Theorem 4.2). *The compound-return Esscher-transformed martingale measure* $P_{Z_{[0,T]}}^{(ESS)}$ *is unique if it exists.*

Proposition 4.2. ([62], Theorem 4.5). *The simple-return Esscher-transformed martingale measure* $P^{(ESS)}_{\hat{Z}_{[0,T]}}$ *is unique if it exists.*

Based on the results stated above, we can give the following definition.

Definition 4.5. (i) The compound-return Esscher-transformed martingale measure $P^{(ESS)}_{Z_{[0,T]}}$ is called the "Esscher martingale measure (ESSMM)", and denoted by $P^{(ESSMM)}$.

(ii) The simple-return Esscher-transformed martingale measure $P^{(ESS)}_{\hat{Z}_{[0,T]}}$ is called the "simple-return Esscher-transformed martingale measure", and denoted by $\hat{P}^{(ESSMM)}$.

Remark 4.2. As we shall see in Chapter 6, the simple-return Esscher-transformed martingale measure $\hat{P}^{(ESSMM)}$ is identified with the minimal entropy martingale measure $P^{(MEMM)}$.

4.3 Existence Theorems of $P^{(ESSMM)}$ and $\hat{P}^{(ESSMM)}$ for Geometric Lévy Processes

The uniqueness theorems are stated in the previous section. We next study the existence problem of Esscher-transformed martingale measures.

4.3.1 *Existence theorem of $P^{(ESSMM)}$*

We suppose that all expectations, which appear in what follows, exist. Then the martingale condition for an Esscher-transformed probability measure $Q = P^{(ESS)}_{Z_{[0,T]},h}$ is given in the following form:

$$E_Q[e^{-r}S_1] = e^{-r}S_0 E_Q[e^{Z_1}] = e^{-r}S_0 \frac{E_P[e^{(h+1)Z_1}]}{E_P[e^{hZ_1}]} = S_0. \qquad (4.15)$$

This condition is equivalent to the following condition:

$$E_P[e^{(h+1)Z_1}] = e^r E_P[e^{hZ_1}], \qquad (4.16)$$

and this is also equivalent to the following condition:

$$\phi(-i(h+1)) = e^r \phi(-ih), \quad \phi(u) = E_P[e^{iuZ_1}], \qquad (4.17)$$

where $\phi(u)$ is the characteristic function of Z_1.

Set the following:

$$f^{(ESSMM)}(h) = b + (\frac{1}{2} + h)\sigma^2 + \int_{\{|x| \le 1\}} \left((e^x - 1)e^{hx} - x\right) \nu(dx)$$

$$+ \int_{\{|x| > 1\}} (e^x - 1)e^{hx} \nu(dx). \tag{4.18}$$

Then we obtain the following theorem:

Theorem 4.1. (Existence condition for ESSMM). *If the equation*

$$f^{(ESSMM)}(h) = r \tag{4.19}$$

has a solution h^*, then the ESSMM of S_t, $P^{(ESSMM)}$, exists and is given by

$$P^{(ESSMM)} = P^{(ESS)}_{Z_{[0,T]}, h^*} = P^{(ESS)}_{Z_T, h^*}. \tag{4.20}$$

The process Z_t is also a Lévy process under $P^{(ESSMM)}$, and the generating triplet of Z_t under $P^{(ESSMM)}$ is given by

$$\sigma^{(ESSMM)^2} = \sigma^2, \tag{4.21}$$

$$\nu^{(ESSMM)}(dx) = e^{h^*x}\nu(dx), \tag{4.22}$$

$$b^{(ESSMM)} = b + h^*\sigma^2 + \int_{\{|x| \le 1\}} x(e^{h^*x} - 1)\nu(dx). \tag{4.23}$$

(Proof) Equation (4.19) is equivalent to (4.17). Therefore $P^{(ESS)}_{Z_{[0,T]}, h^*}$ is a martingale measure of S_t.

The characteristic function of Z_t under $P^{(ESSMM)} = P^{(ESS)}_{Z_{[0,T]}, h^*}$ is, by definition,

$$\phi_t^{(ESSMM)}(u) = E_{P^{(ESSMM)}}[e^{iuZ_t}] = \frac{E_P[e^{iuZ_t}e^{h^*Z_T}]}{E_P[e^{h^*Z_T}]}. \tag{4.24}$$

And this is equal to

$$\frac{E_P[e^{(iu+h^*)Z_t}]}{E_P[e^{h^*Z_t}]} = \frac{\phi_t(u - ih^*)}{\phi_t(-ih^*)}. \tag{4.25}$$

By simple calculation we obtain

$$\phi_t^{(ESSMM)}(u) = \exp\left\{ t \left(-\frac{1}{2}\sigma^2 u^2 \right. \right.$$

$$+ i\left(b + h^*\sigma^2 + \int_{\{|x| \le 1\}} x(e^{h^*x} - 1)\nu(dx)\right) u$$

$$+ \int_{\{|x| \le 1\}} (e^{iux} - 1 - iux)e^{h^*x}\nu(dx)$$

$$\left. \left. + \int_{\{|x| > 1\}} (e^{iux} - 1)e^{h^*x}\nu(dx) \right) \right\}. \tag{4.26}$$

This formula proves the results of the theorem.
(Q.E.D.)

4.3.2   Existence theorem of $\hat{P}^{(ESSMM)}$

Set the following:

$$\hat{f}^{(ESSMM)}(\hat{h}) = b + (\frac{1}{2} + \hat{h})\sigma^2 + \int_{\{|x|\leq 1\}} \left((e^x - 1)e^{\hat{h}(e^x-1)} - x\right)\nu(dx)$$
$$+ \int_{\{|x|>1\}} \left((e^x - 1)e^{\hat{h}(e^x-1)}\right)\nu(dx). \tag{4.27}$$

Then we obtain the following theorem:

Theorem 4.2. (Existence condition for $\hat{P}^{(ESSMM)}$). *If the equation*

$$\hat{f}^{(ESSMM)}(\hat{h}) = r \tag{4.28}$$

has a solution $\hat{h}^*$, then the simple-return Esscher-transformed martingale measure $\hat{P}^{(ESSMM)}$ of S_t exists, and it is given by

$$\hat{P}^{(ESSMM)} = P^{(ESS)}_{\hat{Z}_{[0,T]},\hat{h}^*} = P^{(ESS)}_{\hat{Z}_T,\hat{h}^*}. \tag{4.29}$$

The process Z_t is also a Lévy process under $\hat{P}^{(ESSMM)}$, and the generating triplet of Z_t under $\hat{P}^{(ESSMM)}$ is

$$\hat{\sigma}^{(ESSMM)^2} = \sigma^2, \tag{4.30}$$
$$\hat{\nu}^{(ESSMM)}(dx) = e^{\hat{h}^*(e^x-1)}\nu(dx), \tag{4.31}$$
$$\hat{b}^{(ESSMM)} = b + \hat{h}^*\sigma^2 + \int_{\{|x|\leq 1\}} x(e^{\hat{h}^*(e^x-1)} - 1)\nu(dx). \tag{4.32}$$

(Proof) The martingale condition for $P^{(ESS)}_{\hat{Z}_{[0,T]},\hat{h}}$ is

$$\frac{E[e^{Z_t}e^{\hat{h}\hat{Z}_t}]}{E[e^{\hat{h}\hat{Z}_t}]} = e^{rt}. \tag{4.33}$$

The Lévy processes Z_t and $\hat{Z}_t$ are expressed in the following form:

$$Z_t = \sigma W_t + bt + \int_0^t \int_{|x| \le 1} x \tilde{N}_p(dsdx) + \int_0^t \int_{\{|x| > 1\}} x N_p(dsdx)$$

$$= \sigma W_t + \left(b + \int_{\{|x| > 1\}} x\nu(dx) \right) t + \int_0^t \int_{-\infty}^{\infty} x \tilde{N}_p(dsdx), \quad (4.34)$$

$$\hat{Z}_t = \sigma W_t + \tilde{b}t + \int_0^t \int_{|x| \le 1} (e^x - 1) \tilde{N}_p(dsdx)$$

$$+ \int_0^t \int_{\{|x| > 1\}} (e^x - 1) N_p(dsdx)$$

$$= \sigma W_t + \tilde{b}'t + \int_{0+}^t \int_{-\infty}^{\infty} (e^x - 1) \tilde{N}_p(dsdx), \quad (4.35)$$

where

$$\tilde{b} = b + \frac{1}{2}\sigma^2 + \int_{|x| \le 1} \left((e^x - 1 - x) \right) \nu(dx), \quad (4.36)$$

$$\tilde{b}' = b + \frac{1}{2}\sigma^2 + \int_{-\infty}^{\infty} \left((e^x - 1 - x 1_{\{|x| \le 1\}}(x)) \right) \nu(dx). \quad (4.37)$$

So we obtain

$$Z_t + \hat{h}\hat{Z}_t = (1 + \hat{h})\sigma W_t + (b + \int_{\{|x| > 1\}} x\nu(dx) + \hat{h}\tilde{b}')t$$

$$+ \int_0^t \int_{-\infty}^{\infty} (x + \hat{h}(e^x - 1)) \tilde{N}_p(dsdx). \quad (4.38)$$

Set the following:

$$X_t = Z_t + \hat{h}\hat{Z}_t, \quad (4.39)$$

$$V_t = e^{X_t} = e^{Z_t + \hat{h}\hat{Z}_t}. \quad (4.40)$$

Applying Itô's formula to $F(x) = e^x$, we obtain

$$V_t = \int_0^t V_s \left\{ (1 + \hat{h})\sigma \right\} W_s$$

$$+ \int_0^t V_s \left\{ b + \int_{\{|x| > 1\}} x\nu(dx) + \hat{h}\tilde{b}' + \frac{1}{2}(1 + \hat{h})^2\sigma^2 \right\} ds$$

$$+ \int_0^t \int_{-\infty}^{\infty} V_{s-} \left\{ e^{(x + \hat{h}(e^x - 1))} - 1 \right\} \tilde{N}_p(dsdx)$$

$$+ \int_0^t \int_{-\infty}^{\infty} V_s \left\{ e^{(x + \hat{h}(e^x - 1))} - 1 - (x + \hat{h}(e^x - 1)) \right\} \nu(dx)ds,$$

$$(4.41)$$

and

$$E[V_t] = \int_0^t E[V_s](b + \hat{h}\tilde{b}' + \frac{1}{2}(1+\hat{h})^2\sigma^2)ds$$
$$+ \int_0^t E[V_s] \int_{-\infty}^{\infty} \left\{ e^{(x+\hat{h}(e^x-1))} - 1 \right.$$
$$\left. -(x1_{\{|x|\le 1\}}(x) + \hat{h}(e^x - 1)) \right\} \nu(dx)ds. \qquad (4.42)$$

From this it follows that

$$E[e^{Z_t}e^{\hat{h}\hat{Z}_t}] = E[V_t]$$
$$= \exp\left\{ t\left((b + \hat{h}\tilde{b}' + \frac{1}{2}(1+\hat{h})^2\sigma^2 \right. \right.$$
$$\left. \left. + \int_{-\infty}^{\infty} \left\{ e^{(x+\hat{h}(e^x-1))} - 1 - (x1_{\{|x|\le 1\}}(x) + \hat{h}(e^x - 1)) \right\} \nu(dx) \right) \right\}.$$
$$(4.43)$$

On the other hand, it is easy to see that

$$E[e^{\hat{h}\hat{Z}_t}] = \phi_{\hat{Z}}(-i\hat{h})$$
$$= \exp\left\{ t\left((\hat{h}\hat{b} + \frac{1}{2}\hat{h}^2\sigma^2) \right. \right.$$
$$\left. \left. + \int_{-\infty}^{\infty} \left(e^{\hat{h}(e^x-1)} - 1 - \hat{h}(e^x - 1)1_{\{e^x-1\le 1\}}(x) \right) \nu(dx) \right) \right\}, \quad (4.44)$$

where

$$\hat{b} = b + \frac{1}{2}\sigma^2 + \int_{-\infty}^{\infty} \left((e^x - 1)1_{\{e^x-1\le 1\}} - x1_{\{|x|\le 1\}}(x) \right) \nu(dx). \qquad (4.45)$$

Therefore we have obtained

$$\frac{E[e^{Z_t}e^{\hat{h}\hat{Z}_t}]}{E[e^{\hat{h}\hat{Z}_t}]}$$
$$= \exp\left\{ t\left((b + (\frac{1}{2} + \hat{h})\sigma^2) + \int_{\{|x|\le 1\}} \left((e^x - 1)e^{\hat{h}(e^x-1)} - x \right) \nu(dx) \right. \right.$$
$$\left. \left. + \int_{\{|x|>1\}} \left((e^x - 1)e^{\hat{h}(e^x-1)} \right) \nu(dx) \right) \right\}$$
$$= e^{t\hat{f}^{(ESSMM)}(\hat{h})}. \qquad (4.46)$$

From (4.46) and the assumption that $\hat{h}^*$ is a solution of (4.28), it follows that

$$\frac{E[e^{Z_t}e^{\hat{h}^*\hat{Z}_t}]}{E[e^{\hat{h}^*\hat{Z}_t}]} = e^{t\hat{f}^{(ESSMM)}(\hat{h}^*)} = e^{rt}. \tag{4.47}$$

This equality proves that the martingale condition (4.33) is satisfied by $\hat{h}^*$. So it is proved that $P^{(ESS)}_{\hat{Z}_{[0,T]},\hat{h}^*}$ is a martingale measure.

We next calculate the Radon–Nikodým derivative of $\hat{P}^{(ESSMM)}$ with respect to P. It is easily proved that

$$\frac{d\hat{P}^{(ESSMM)}}{dP}\Big|_{\mathcal{F}_t} = \frac{e^{\hat{h}^*\hat{Z}_t}}{E[e^{\hat{h}^*\hat{Z}_t}]}$$

$$= \frac{e^{\left\{\hat{h}^*\left(\sigma W_t + \tilde{b}'t + \int_0^t\int_{-\infty}^{\infty}(e^x-1)\tilde{N}_p(dsdx)\right)\right\}}}{e^{\left\{t\left((\hat{h}^*\hat{b}+\frac{1}{2}\hat{h}^{*2}\sigma^2)+\int_{-\infty}^{\infty}\left(e^{\hat{h}^*(e^x-1)}-1-\hat{h}^*(e^x-1)1_{\{e^x-1\leq 1\}}(x)\right)\nu(dx)\right)\right\}}}$$

$$= e^{\left\{\left(\hat{h}^*\sigma W_t + \hat{h}^*\tilde{b}'t\right)-\left((\hat{h}^*\hat{b}+\frac{1}{2}\hat{h}^{*2}\sigma^2)t\right)\right\}}$$

$$\times e^{\left\{\int_0^t\int_{-\infty}^{\infty}\hat{h}^*(e^x-1)\tilde{N}_p(dsdx)-\left(\int_{-\infty}^{\infty}\left(e^{\hat{h}^*(e^x-1)}-1-\hat{h}^*(e^x-1)1_{\{e^x-1\leq 1\}}(x)\right)\nu(dx)\right)t\right\}}$$

$$= \exp\left\{\hat{h}^*\sigma W_t - \frac{1}{2}\hat{h}^{*2}\sigma^2 t + \int_0^t\int_{-\infty}^{\infty}\hat{h}^*(e^x-1)\tilde{N}_p(dsdx)\right.$$

$$\left. - \left(\int_{-\infty}^{\infty}\left(e^{\hat{h}^*(e^x-1)}-1-\hat{h}^*(e^x-1)\right)\nu(dx)\right)t\right\}, \tag{4.48}$$

where we used (4.36) and (4.37).

The characteristic function of Z_t with respect to $\hat{P}^{(ESSMM)}$ is

$$\hat{\phi}_t^{(ESSMM)}(u) = E_{\hat{P}^{(ESSMM)}}[e^{iuZ_t}] = E_P\left[e^{iuZ_t}\frac{d\hat{P}^{(ESSMM)}}{dP}\Big|_{\mathcal{F}_t}\right]. \tag{4.49}$$

Using (4.48), we can do similar calculations as we have done above for $E[e^{Z_t}e^{\hat{h}\hat{Z}_t}]$, and we obtain

$$\hat{\phi}_t^{(ESSMM)}(u) = E_{\hat{P}^{(ESSMM)}}\left[e^{iuZ_t}\right]$$

$$= \exp\left\{t\left(-\frac{1}{2}\sigma^2u^2 + i\left(b+\hat{h}^*\sigma^2+\int_{\{|x|\leq 1\}}x\left(e^{\hat{h}^*(e^x-1)}-1\right)\nu(dx)\right)u\right.\right.$$

$$+\int_{\{|x|\leq 1\}}(e^{iux}-1-iux)e^{\hat{h}^*(e^x-1)}\nu(dx)$$

$$\left.\left.+\int_{\{|x|>1\}}(e^{iux}-1)e^{\hat{h}^*(e^x-1)}\nu(dx)\right)\right\}. \tag{4.50}$$

This formula proves that the generating triplet of Z_t under $\hat{P}^{(ESSMM)}$ is of the form stated in the theorem.
(Q.E.D.)

4.4 Comparison of $P^{(ESSMM)}$ and $\hat{P}^{(ESSMM)}$

As we have mentioned in Remark 4.2, the simple-return Esscher-transformed martingale measure $\hat{P}^{(ESSMM)}$ is identified with the minimal entropy martingale measure $P^{(MEMM)}$. So, from now on, we will use the notation $P^{(MEMM)}$ identically with $\hat{P}^{(ESSMM)}$.

Both the ESSMM, $P^{(ESSMM)}$, and the MEMM, $P^{(MEMM)} = \hat{P}^{(ESSMM)}$, are obtained by Esscher transformation, but they have different properties. We shall survey the differences between them.

1) For the existence of $P^{(ESSMM)}$, the following condition

$$\int_{\{|x|>1\}} |(e^x - 1)e^{h^* x}|\, \nu(dx) < \infty \qquad (4.51)$$

is necessary, where h^* is a constant which is determined by (4.19) in Theorem 4.1. On the other hand, for the existence of $P^{(MEMM)}$, the corresponding condition is

$$\int_{\{|x|>1\}} |(e^x - 1)e^{\hat{h}^*(e^x - 1)}|\, \nu(dx) < \infty, \qquad (4.52)$$

where $\hat{h}^*$ is a constant which is determined by (4.28) in Theorem 4.2. Condition (4.52) is satisfied for a wide class of Lévy measures when $\hat{h}^* < 0$. Namely, the former condition is strictly stronger than the latter condition. This means that the MEMM, $P^{(MEMM)} = \hat{P}^{(ESSMM)}$, can be applied to a wider class of models rather than the ESSMM, $P^{(ESSMM)}$. An important example of this difference appears in the case of stable process models. In fact we can apply $\hat{P}^{(ESSMM)} = P^{(MEMM)}$ to geometric stable process models but we can not apply $P^{(ESSMM)}$ to geometric stable process models. (See Chapters 7 and 9.)

2) The relative entropy is very popular in the field of information theory, and it is called Kullback–Leibler information number (see [56] p.23) or Kullback–Leibler distance (see [27] p.18). Therefore we can state that the MEMM is the nearest equivalent martingale measure to the original probability P in the sense of Kullback–Leibler distance. The idea of minimal

distance martingale measure is mentioned in [47], and relative entropy is a typical example of it. We shall see these subjects in Chapter 6.

3) Large deviation theory is closely related to the minimum relative entropy analysis, and Sanov's theorem or the Sanov property is well known (see, for example, [27] pp.291–304 or [56] pp.110–111). This theorem says that the MEMM is the most possible empirical probability measure of paths of price process in the class of equivalent martingale measures. In this sense the MEMM is considered to be an exceptional measure in the class of all equivalent martingale measures.

4.5   Other Examples of Esscher-Transformed Martingale Measures

If we adopt a different risk process than Z_t or $\hat{Z}_t$, then we obtain another kind of Esscher-transformed martingale measure.

For jump-diffusion models, the Brownian motion can be adopted as a risk process. In such cases the corresponding Esscher-transformed martingale measure is the mean correcting martingale measure (MCMM). Or we can adopt the jump part of a jump-diffusion process as a risk process. (See [87] or [19].)

Notes

The argument of this chapter is based on the results of Miyahara (2004) [89], and contains some extensions.

The Esscher transformation was first introduced by Esscher (1932) [35], and has been developed in risk management theory (see Bühlmann (1970) [11] for example). The Esscher martingale measure (ESSMM) is very popular nowadays, but there are only few articles that investigate Esscher transformation from the theoretical point of view. Among them, Bühlmann, Delbaen, Embrechts, and Shiryaev (1996) [12] and Kallsen and Shiryaev (2002) [62] are notable.

Relations between the ESSMM and the MEMM for the geometric Lévy process models are investigated in Fujiwara and Miyahara (2003) [44] and Hubalek and Sgarra (2006) [53].

Chapter 5

Minimax Martingale Measures and Minimal Distance Martingale Measures

In this chapter we investigate martingale measures based on utility functions (U-MM), and we also see that such martingale measures correspond to minimal distance martingale measures (MDMM).

5.1 Utility Function, Duality, and Minimax Martingale Measures

We give the definition of a utility function as follows.

Definition 5.1. A function $u(x)$ defined on $(-\infty, \infty)$ or $(0, \infty)$ is called a utility function if it is strictly increasing, strictly concave, and continuously differentiable.

Remark 5.1. Sometimes we assume that a utility function is of $C^{(2)}$-class.

Definition 5.2. The convex conjugate function $u^*(y)$ of a utility function $u(x)$ is given by the following formula:

$$u^*(y) = \sup_x \left(u(x) - xy\right) = u(I(y)) - yI(y), \qquad (5.1)$$

where

$$I(y) = (u')^{-1}(y). \qquad (5.2)$$

We suppose that a probability space $(\Omega, \mathcal{F}, P)$ and a filtration $\{\mathcal{F}_t\}$ are given as usual. And let S_t be an underlying asset process defined on this probability space.

Set the following:

$$\mathcal{M}_1 = \{Q \ll P : E_Q[w] \le 0, \quad \forall w \in C\}, \qquad (5.3)$$

where C is a convex cone in L^∞ which is defined related to the process S_t. (See [6] pp.2-3 for the precise definition of C and $\mathcal{M}_1$.) It is known that if S_t is bounded then $\mathcal{M}_1$ is the set of all equivalent martingale measures of S_t. (See [6] p.4.)

For $Q \in \mathcal{M}_1$ and $x \in (-\infty, \infty)$, set

$$U_P(x; Q) = \sup\{E_P[u(Y)] : Y \in L^\infty, E_Q[Y] \le x\} \qquad (5.4)$$

and

$$U_P(x) = \sup\{E_P[u(Y)] : Y \in L^\infty, E_Q[Y] \le x \quad \forall Q \in \mathcal{M}_1\}. \qquad (5.5)$$

$U_P(x; Q)$ is called *Q-utility* for x. From the definitions above it follows that

$$U_P(x) \le U_P(x; Q). \qquad (5.6)$$

Definition 5.3. An equivalent probability measure $Q^*(x) \in \mathcal{M}_1$ is called a minimax measure for x if the following conditions are satisfied:

$$U_P(x; Q^*(x)) = \inf\{U_P(x; Q); Q \in \mathcal{M}_1\} \qquad (5.7)$$

and

$$U_P(x) = U_P(x; Q^*(x)). \qquad (5.8)$$

Definition 5.4. Let $Q^*(x)$ be a minimax measure for x.
(i) For a contingent claim (i.e. a $\mathcal{F}_T$-adaptive random variable) X, $E_{Q^*(x)}[X]$ is called the minimax price of X for x.
(ii) If $Q^*(x)$ is a martingale measure of S_t, then $Q^*(x)$ is called the minimax martingale measure of S_t for x.

Remark 5.2. $Q^*(x)$ depends on x in general, but in some cases it is independent of x.

The economic implication of the minimax pricing is that it "produces prices which are least favorable for an investor with a given utility profile, i.e., the maximal expected utility with respect to prices based on a martingale measure is minimal". (See [47] p.564.) The relation of minimax pricing with Arrow–Debreu state prices is disscussed in [51] and [52].

5.2 Distance Function Corresponding to Utility Function

The problem in obtaining a minimax martingale measure for a given utility function $u(x)$ is equivalent to the problem in obtaining a minimal distance martingale measure for the dual distance function $u^*(y)$. The key theorem is the following proposition. (See [47] Lemma 4.1 or [6] Corollary 2.1 and p.17.)

Proposition 5.1. *Assume that $Q \ll P$ and $E_Q[I(\lambda \frac{dQ}{dP})] < \infty$ for all $\lambda > 0$. Then it holds that*

$$U_P(x; Q) = E_P[u(I(\lambda_Q(x)\frac{dQ}{dP}))], \qquad (5.9)$$

where $\lambda_Q(x)$ is determined by

$$E_Q[I(\lambda \frac{dQ}{dP})] = x. \qquad (5.10)$$

Using Proposition 5.1, we can find a distance function corresponding to a utility function.

Let us first see the case of exponential utility function. Set the following:

$$u(x) = u_\alpha(x) = \frac{1}{\alpha}\left(1 - e^{-\alpha x}\right), \quad \alpha > 0, \qquad (5.11)$$

then we have

$$I(y) = (u')^{-1}(y) = -\frac{1}{\alpha}\log y, \qquad (5.12)$$

and

$$u(I(y)) = \frac{1}{\alpha}(1 - y). \qquad (5.13)$$

The conjugate function is

$$u^*(y) = u_\alpha^*(y) = \frac{1}{\alpha}(y\log y + (1 - y)). \qquad (5.14)$$

Equation (5.10) for $\lambda_Q(x)$ is

$$E_Q[I(\lambda \frac{dQ}{dP})] = -\frac{1}{\alpha}\left(\log\lambda + E_Q[\log\frac{dQ}{dP}]\right) = x, \qquad (5.15)$$

and the solution is

$$\lambda_Q(x) = e^{-E_P[\frac{dQ}{dP}\log\frac{dQ}{dP}]-\alpha x} = e^{-H(Q|P)-\alpha x}, \qquad (5.16)$$

where $H(Q|P)$ is the relative entropy of Q with respect to P. From (5.9) and (5.13) we obtain

$$U_P(x; Q) = E_P\left[u(I(\lambda_Q(x)\frac{dQ}{dP}))\right] = E_P\left[\frac{1}{\alpha}\left(1 - e^{-H(Q|P)-\alpha x}\frac{dQ}{dP}\right)\right]$$

$$= \frac{1}{\alpha}\left(1 - e^{-H(Q|P)-\alpha x}\right). \qquad (5.17)$$

From this formula it follows that the minimax martingale measure $Q^*(x)$ is a solution of the minimization problem of $H(Q|P)$. We notice that in the case of exponential utility function $Q^*(x)$ does not depend on x.

Thus, setting $F(x) = x \log x$, we have a problem finding the minimal distance martingale measure for the distance function $F(x)$:

$$E_P \left[F(\frac{dQ^*}{dP}) \right] = \inf_{Q \in \mathcal{M}_1} \{E_P \left[F(\frac{dQ}{dP}) \right]\}. \tag{5.18}$$

In this case the solution is the minimal entropy martingale measure (MEMM).

In the same manner, we can calculate other cases, and obtain the following examples of distance function. (See [6] or [47].)

(1) Utility function: $u_\alpha(x) = \frac{1}{\alpha}(1 - e^{-\alpha x})$, $\alpha > 0$,
   Conjugate function: $u_\alpha^*(y) = \frac{1}{\alpha}(y \log y + (1 - y))$,
   Q-utility: $U_P(x; Q) = \frac{1}{\alpha} \left(1 - e^{-H(Q|P) - \alpha x} \right)$,
   Distance function: $F(x) = x \log x$.

(2) Utility function: $u(x) = \log x$,
   Conjugate function: $u^*(y) = -\log y - 1$,
   Q-utility: $U_P(x; Q) = \log x - E_P[\log \frac{dQ}{dP}] = \log x + E_Q[\frac{dP}{dQ} \log \frac{dP}{dQ}] = \log x + H(P|Q)$,
   Distance function: $F(x) = -\log x$.

(3) Utility function: $u(x) = \frac{1}{p}x^p, x \in (0, \infty)$, $\quad p < 1, p \neq 0$,
   Conjugate function: $u^*(y) = \frac{-1}{q}y^q, y \in (0, \infty)$, $\quad q = \frac{p}{p-1}$,
   Q-utility: $U_P(x; Q) = u(x)E_P[(\frac{dQ}{dP})^q]^{\frac{-1}{q-1}}, q = \frac{p}{p-1}$,
   Distance function: $F(x) = \frac{-1}{q}x^q, x \in (0, \infty)$.

   1) If $p < 0$, then $0 < q = \frac{p}{p-1} < 1$.
   2) If $0 < p < 1$, then $-\infty < q = \frac{p}{p-1} < 0$.

Remark 5.3. In the case of $p > 1$, $\frac{1}{p}x^p$ is not concave. If we put $u(x) = \frac{-1}{p}x^p$, then $u(x)$ is concave, and the formal conjugate function is $u^*(y) = \frac{1}{q}y^q$, $q = \frac{p}{p-1} > 1$, and we obtain a formal distance function $F(x) = \frac{1}{q}x^q, q > 1$.

5.3   Minimal Distance Martingale Measures

Suppose that a utility function $u(x)$ is given and let $F(x)$ be the corresponding distance function. If there exists an equivalent martingale measure Q^* which satisfies the condition

$$E_P[F(\frac{dQ^*}{dP})] = \inf_{Q \in \mathcal{M}_1} \{E_P[F(\frac{dQ}{dP})]\}, \qquad (5.19)$$

then Q^* is the minimal distance martingale measure (MDMM) for a distance function $F(x)$. This Q^* is also the minimax martingale measure for the utility function $u(x)$.

The following names are given to special cases:

(1) $u(x) = 1 - e^{-\alpha x} \to F(x) = x \log x$,
   "minimal entropy martingale measure (MEMM)".

(2) $u(x) = \log x \to F(x) = -\log x$,
   "minimal reverse relative entropy martingale measure".

(3) $u(x) = \frac{1}{p} x^p$, $p \neq 0, 1$ (If $p > 1$, then $u(x) = \frac{-1}{p}|x|^p$),
   $\to F(x) = \frac{-1}{q} x^q, -\infty < q = \frac{p}{p-1} < \infty, q \neq 0, 1$ (If $p > 1$, then $F(x) = \frac{1}{q}|x|^q$).
   "minimal q-moment martingale measure (ML^{q}MM)".

   (3-1) $p = -1 \to F(x) = -\frac{1}{4}\sqrt{x}$,
   "minimal Hellinger distance martingale measure",

   (3-2) $p = 2 \to F(x) = \frac{1}{2}|x|^2$,
   "minimal variance equivalent martingale measure (MVEMM)".

In Chapter 6 we shall study the minimal distance martingale measures corresponding to the distance functions stated above.

Notes

The statements of this chapter relate to the following papers:

Bellini and Frittelli (2002) [6], Goll and Rüschendorf (2001) [47], and Gundel (2005) [50].

Utility based martingale measures (U-MM) are related to the expected utility maximization problem and the utility indifference pricing theory. In such fields the duality method is very efficient. There are many papers and books on this subject. Among them the following papers and books are notable:

Kramkov and Schachermayer (1999) [69], Delbaen, Grandits, Rheinländer, Samperi, Schweizer, and Stricker (2002) [29], Kallsen (2002) [61], and Carmona (Editor) (2009) [13].

Chapter 6

Minimal Distance Martingale Measures for Geometric Lévy Processes

As we have seen in Chapter 5, a minimal distance martingale measure appears when we study the utility function based martingale measure. This is the second method to construct martingale measures. (The first one is the Esscher transformation method.)

6.1 Minimal Distance Problem

Let Z_t be a Lévy process on a probability space $(\Omega, \mathcal{F}, P)$, and let (σ^2, ν, b) be its generating triplet. Then the Lévy–Itô decomposition of Z_t is

$$
\begin{aligned}
Z_t &= \sigma W_t + bt + \int_0^t \int_{\{|x|>1\}} x N(ds, dx) + \int_0^t \int_{\{|x|\leq 1\}} x \widetilde{N}(ds, dx) \\
&= \sigma W_t + bt + \int_0^t \int_{-\infty}^{\infty} x \widetilde{N}(ds, dx) + \int_0^t \int_{\{|x|>1\}} x\nu(dx)ds, \quad (6.1)
\end{aligned}
$$

where W_t is a Brownian motion, $N(ds, dx)$ is a Poisson random measure, and $\widetilde{N}(dt, dx) = N(dt, dx) - dt\nu(dx)$. We suppose that the price of a risky asset is $S_t = S_0 e^{Z_t}$, and that the risk-free asset has a constant interest rate r. The discounted process of S_t is $\widetilde{S}_t = S_t e^{-rt}$.

For a probability measure Q equivalent to P, set

$$
L_t^Q = \frac{dQ}{dP}\big|_{\mathcal{F}_t}. \quad (6.2)
$$

Then $L_t^Q, t \geq 0$ is a P-martingale. From the predictable representation theorem (see [70] Theorem 2.1) there exist two predictable processes, f_t and $g_t = g(t, x)$, such that

$$
dL_t^Q = L_{t-}^Q \left(f_t dW_t + \int_{-\infty}^{\infty} (e^{g(t,x)} - 1) \widetilde{N}(dt, dx) \right). \quad (6.3)
$$

Conversely, for any pair (f, g), the equation (6.3) defines a strictly positive martingale $L_t(f, g)$. We denote by $Q^{L(f,g)}$ the corresponding equivalent probability measure which is defined by

$$\frac{dQ^{L(f,g)}}{dP}|_{\mathcal{F}_t} = L_t(f, g). \tag{6.4}$$

When $\int_0^t \int_{-\infty}^{\infty} g(s, x) \, N(dt, dx)$ is well defined, the process $L_t(f, g)$ can be expressed in the following form:

$$
\begin{aligned}
L_t(f, g) &= \exp\left\{ \int_0^t f_s dW_s - \frac{1}{2} \int_0^t f_s^2 ds + \int_0^t \int_{-\infty}^{\infty} g(s, x) \, N(ds, dx) \right. \\
&\quad \left. - \int_0^t \int_{-\infty}^{\infty} \left(e^{g(s,x)} - 1 \right) \nu(dx) ds \right\} \\
&= \exp\left\{ \int_0^t f_s dW_s - \frac{1}{2} \int_0^t f_s^2 ds + \int_0^t \int_{-\infty}^{\infty} g(s, x) \, \widetilde{N}(ds, dx) \right. \\
&\quad \left. - \int_0^t \int_{-\infty}^{\infty} \left(e^{g(s,x)} - 1 - g(s, x) \right) \nu(dx) ds \right\}. \tag{6.5}
\end{aligned}
$$

For simplicity we shall discuss only such cases where $L_t(f, g)$ is expressed as above, and we restrict our attention to processes L_t which are square integrable.

We investigate the process Z_t under an equivalent probability measure $Q = Q^{L(f,g)}$. The characteristic function of Z_t with respect to Q is

$$E_Q[e^{iuZ_t}] = E_P[e^{iuZ_t} L_t(f, g)] = E_P[e^{X_t}], \tag{6.6}$$

where X_t is given by

$$
\begin{aligned}
X_t &= iuZ_t + \log L_t(f, g) \\
&= \int_0^t (iu\sigma + f_s) dW_s + \int_0^t (iub - \frac{1}{2} f_s^2) ds \\
&\quad + \int_0^t \int_{-\infty}^{\infty} (iux + g(s, x)) \, \widetilde{N}(ds, dx) \\
&\quad - \int_0^t \int_{-\infty}^{\infty} \left(e^{g(s,x)} - 1 - g(s, x) - iux \mathbf{1}_{\{|x|>1\}}(x) \right) \nu(dx) ds. \tag{6.7}
\end{aligned}
$$

Applying Itô's formula to the function e^x, we obtain

$$
\begin{aligned}
e^{X_t} &= \int_0^t e^{X_s}(iu\sigma + f_s)dW_s + \int_0^t e^{X_s}\left(\frac{1}{2}(iu\sigma + f_s)^2 + (iub - \frac{1}{2}f_s^2)\right)ds \\
&+ \int_0^t \int_{-\infty}^{\infty} e^{X_{s-}}\left(e^{g(s,x)+iux} - 1\right)\widetilde{N}(ds, dx) \\
&+ \int_0^t \int_{-\infty}^{\infty} e^{X_{s-}}\left(e^{g(s,x)+iux} - 1 - g(s,x) - iux\right)\nu(dx)ds \\
&- \int_0^t \int_{-\infty}^{\infty} e^{X_{s-}}\left(e^{g(s,x)} - 1 - g(s,x) - iux1_{\{|x|>1\}}(x)\right)\nu(dx)ds \\
&= \int_0^t e^{X_s}(iu\sigma + f_s)dW_s + \int_0^t \int_{-\infty}^{\infty} e^{X_{s-}}\left(e^{g(s,x)+iux} - 1\right)\widetilde{N}(ds, dx) \\
&+ \int_0^t e^{X_{s-}}\left\{-\frac{1}{2}\sigma^2 u^2 + i(b + \sigma f_s)u\right. \\
&\left.+ \int_{-\infty}^{\infty}\left(e^{g(t,x)}(e^{iux} - 1) - iux1_{\{|x|\leq 1\}}(x)\right)\nu(dx)\right\}ds, \quad (6.8)
\end{aligned}
$$

and so we have

$$
\begin{aligned}
E_P[e^{X_t}] &\\
= E_P&\left[\int_0^t e^{X_{s-}}\left\{-\frac{1}{2}\sigma^2 u^2 + i(b + \sigma f_s)u\right.\right. \\
&\left.\left.+ \int_{-\infty}^{\infty}\left(e^{g(t,x)}(e^{iux} - 1) - iux1_{\{|x|\leq 1\}}(x)\right)\nu(dx)\right\}ds\right]. \quad (6.9)
\end{aligned}
$$

In this formula we consider the cases where the pair $(f_s, g(s, x))$ is deterministic and does not depend on s. Then we obtain the following formula:

$$
\begin{aligned}
E_P[e^{X_t}] &\\
= \int_0^t& E_P[e^{X_{s-}}]\left\{-\frac{1}{2}\sigma^2 u^2 + i(b + \sigma f)u\right. \\
&\left.+ \int_{-\infty}^{\infty}\left(e^{g(x)}(e^{iux} - 1) - iux1_{\{|x|\leq 1\}}(x)\right)\nu(dx)\right\}ds. \quad (6.10)
\end{aligned}
$$

Solving this equation for $E_P[e^{X_t}]$, we obtain

$$
\begin{aligned}
&E_P[e^{X_t}] \\
&= \exp\left\{ \int_0^t \left\{ -\frac{1}{2}\sigma^2 u^2 + i(b+\sigma f)u \right.\right. \\
&\qquad\qquad \left.\left. + \int_{-\infty}^{\infty} \left(e^{g(x)}(e^{iux}-1) - iux1_{\{|x|\le 1\}}(x) \right) \nu(dx) \right\} ds \right\} \\
&= \exp\left\{ t \left\{ -\frac{1}{2}\sigma^2 u^2 + i(b+\sigma f)u \right.\right. \\
&\qquad\qquad \left.\left. + \int_{-\infty}^{\infty} \left(e^{g(x)}(e^{iux}-1) - iux1_{\{|x|\le 1\}}(x) \right) \nu(dx) \right\} \right\} \\
&= \exp\left\{ t \left\{ -\frac{1}{2}\sigma^2 u^2 + i \left((b+\sigma f) + \int_{-\infty}^{\infty} x1_{\{|x|\le 1\}}(x)(e^{g(x)}-1)\nu(dx) \right) u \right.\right. \\
&\qquad\qquad \left.\left. + \int_{-\infty}^{\infty} \left((e^{iux}-1-iux1_{\{|x|\le 1\}}(x)) \, e^{g(x)}\nu(dx) \right) \right\} \right\}. \quad (6.11)
\end{aligned}
$$

Thus we have obtained a result which shows that the characteristic function of Z_t with respect to $Q = Q^{L(f,g)}$ is

$$
\begin{aligned}
E_Q[e^{iuZ_t}] &= E_P[e^{X_t}] \\
&= \exp\left\{ t \left\{ -\frac{1}{2}\sigma^2 u^2 + i \left((b+\sigma f) + \int_{-\infty}^{\infty} x1_{\{|x|\le 1\}}(x)(e^{g(x)}-1)\nu(dx) \right) u \right.\right. \\
&\qquad\qquad \left.\left. + \int_{-\infty}^{\infty} \left((e^{iux}-1-iux1_{\{|x|\le 1\}}(x)) \, e^{g(x)}\nu(dx) \right) \right\} \right\}. \quad (6.12)
\end{aligned}
$$

From this formula we know that the Lévy measure $\nu^Q(dx)$ of Z_t under $Q = Q^{L(f,g)}$ is

$$
\nu^Q(dx) = e^{g(x)}\nu(dx). \quad (6.13)
$$

Next we investigate the martingale property of an underlying process under $Q^{L(f,g)}$. Suppose that an underlying process is given by $S_t = S_0 e^{Z_t}$. Then the martingale property of S_t is that $\widetilde{S}_t = e^{-rt}S_t$ is martingale under $Q^{L(f,g)}$. We can show the following proposition:

Proposition 6.1. *The process* $\widetilde{S}_t = e^{-rt}S_t = e^{-rt+Z_t}$ *is martingale under* $Q^{L(f,g)}$ *if and only if for any t the equality*

$$
b + \frac{1}{2}\sigma^2 + f_t\sigma + \int_R (e^{g(t,x)}(e^x-1) - x1_{\{|x|\le 1\}}(x))\nu(dx) = r \quad (6.14)
$$

holds almost surely.

(Proof) (We give only a rough proof of this proposition. See [70] Theorem 3.1 for the concrete proof.)

We first remark that

$$E_{Q^{L(f,g)}}[\widetilde{S}_t] = S_0 E_P[e^{Z_t - rt} L_t(f,g)] = S_0 E_P[e^{Y_t}], \qquad (6.15)$$

where

$$Y_t = \int_0^t (\sigma + f_s) dW_s + \int_0^t (b - \frac{1}{2} f_s^2 - r) ds$$

$$+ \int_0^t \int_{-\infty}^\infty (g(s,x) + x) \widetilde{N}(ds, dx)$$

$$- \int_0^t \int_{-\infty}^\infty \left(e^{g(s,x)} - 1 - g(s,x) - x 1_{\{|x|>1\}}(x) \right) \nu(dx) ds. \quad (6.16)$$

From this we know that

$$\widetilde{S}_t \text{ is a } Q^{L(f,g)} \text{ martingale}$$
$$\Leftrightarrow e^{Y_t} \text{ is a } P \text{ martingale}. \qquad (6.17)$$

Applying Itô's formula to the function e^x, we obtain

$$e^{Y_t} = \int_0^t e^{Y_s} (\sigma + f_s) dW_s + \int_0^t e^{Y_s} \left(\frac{1}{2} (\sigma + f_s)^2 + (b - \frac{1}{2} f_s^2 - r) \right) ds$$

$$+ \int_0^t \int_{-\infty}^\infty e^{Y_{s-}} \left(e^{g(s,x)+x} - 1 \right) \widetilde{N}(ds, dx)$$

$$+ \int_0^t \int_{-\infty}^\infty e^{Y_{s-}} \left(e^{g(s,x)+x} - 1 - g(s,x) - x \right) \nu(dx) ds$$

$$- \int_0^t \int_{-\infty}^\infty e^{Y_{s-}} \left(e^{g(s,x)} - 1 - g(s,x) - x 1_{\{|x|>1\}}(x) \right) \nu(dx) ds$$

$$= (P - martingale\ part)$$

$$+ \int_0^t e^{Y_{s-}} \left\{ \left((b + \frac{1}{2} \sigma^2 - r) + \sigma f_s \right) \right.$$

$$\left. + \int_{-\infty}^\infty \left(e^{g(t,x)}(e^x - 1) - x 1_{\{|x|\le 1\}}(x) \right) \nu(dx) \right\} ds. \qquad (6.18)$$

Therefore we have obtained the result (6.14).
(Q.E.D.)

We set the following:

$$\mathcal{C} = \{(f,g);\ satisfies\ the\ martingale\ condition (6.14)\}. \qquad (6.19)$$

Each element (f, g) of $\mathcal{C}$ is related to a martingale measure $Q^{L(f,g)}$, where $L(f, g)$ is given by (6.5).

Suppose that a C^2 function $F(x)$ is given and fixed. We restrict our attention to processes $L_t(f, g)$ for which $F(L_t)$ is integrable. For a given function $F(x)$, we set

$$\mathcal{C}_F = \{(f, g) \in \mathcal{C} \text{ and } F(L_t) \text{ is integrable}\}. \tag{6.20}$$

Definition 6.1. The minimal distance martingale measure (MDMM) for a distance function $F(x)$ is the equivalent probability measure Q_F, which satisfies the following conditions:

$$\frac{dQ_F}{dP}\Big|_{\mathcal{F}_T} = L_T(f^*, g^*), \tag{6.21}$$

where (f^*, g^*) is a pair in $\mathcal{C}_F$, which satisfies the following condition:

$$E_P[F(L_T(f^*, g^*))] = \inf_{(f,g)\in\mathcal{C}_F} \{E_P[F(L_T(f, g))]\}. \tag{6.22}$$

(This is the equivalent martingale version of the "minimal distance martingale measure" (MDMM) or "f-projection" in p.559 of Goll and Rüschendorf [47].)

From Itô's formula, we obtain for $F \in C^2$:

$$
\begin{aligned}
& dF(L_t) \\
&= L_t f_t F'(L_t) dW_t + \frac{1}{2} L_t^2 f_t^2 F''(L_t) dt \\
&\quad + \int_{-\infty}^{\infty} \Big(F(L_{t-} + L_{t-}(e^{g(t,x)} - 1)) - F(L_{t-}) \Big) \widetilde{N}(dt, dx) \\
&\quad + \int_{-\infty}^{\infty} \Big(F(L_{t-} + L_{t-}(e^{g(t,x)} - 1)) - F(L_{t-}) \\
&\qquad\qquad\qquad - L_{t-}(e^{g(t,x)} - 1)F'(L_{t-}) \Big) \nu(dx) dt \\
&= L_t f_t F'(L_t) dW_t + \frac{1}{2} L_t^2 f_t^2 F''(L_t) dt \\
&\quad + \int \Big(F(L_{t-} e^{g(t,x)}) - F(L_{t-}) \Big) \widetilde{N}(dt, dx) \\
&\quad + \int_{-\infty}^{\infty} \Big(F(L_{t-} e^{g(t,x)}) - F(L_{t-}) - L_{t-}(e^{g(t,x)} - 1)F'(L_{t-}) \Big) \nu(dx) dt,
\end{aligned}
$$
$$\tag{6.23}$$

and so, assuming integrability, we obtain

$$E[F(L_t)]$$
$$= F(1) + \int_0^t E[L_s^2 f_s^2 F''(L_s)]ds$$
$$+ \int_0^t \left(\int_{-\infty}^\infty E[F(L_s e^{g(s,x)}) - F(L_s) - L_s(e^{g(s,x)} - 1)F'(L_s)]\nu(dx) \right) ds.$$
$$(6.24)$$

This formula is very useful, and our discussions below are based on this formula.

In the following sections we see examples of distance functions and the corresponding minimal distance martingale measures.

6.2 The Minimal Variance Equivalent Martingale Measure (MVEMM)

Let us consider the case where the given distance function is $F(x) = x^2$. Then our aim is to find a pair of predictable processes $(f_t^*, g^*(t,x)) \in \mathcal{C}$ such that

$$E[L_T^2(f^*, g^*)] = \inf_{(f,g)\in\mathcal{C}}\{E[L_T^2(f,g)]\}. \qquad (6.25)$$

If there exists such a pair (f^*, g^*), then the corresponding probability $Q^{L(f^*,g^*)}$ is the minimal variance equivalent martingale measure (MVEMM).

Remark 6.1. (1) We treat the problem in the class of "equivalent martingale measures". (Note that this class is the most desirable class for the option pricing theory, compared to the class of signed martingale measures or the class of absolutely continuous martingales.)

(2) The variance optimal martingale measure (VOMM) is not always the MVEMM. When a VOMM is proved to be positive and equivalent to the original probability P, then the VOMM is the MVEMM.

In the case of $F(x) = x^2$, the equations (6.23) and (6.24) are

$$dL_t^2 = 2L_{t-}^2 \left(f_t dW_t + \int_{-\infty}^\infty (e^{2g} - 1)\widetilde{N}(dt, dx) \right)$$
$$+ L_t^2 \left(f_t^2 + \int_{-\infty}^\infty (e^{g(t,x)} - 1)^2 \nu(dx) \right) dt \qquad (6.26)$$

and

$$E[L_t^2] = 1 + \int_0^t E\left[L_s^2\left(f_s^2 + \int_{-\infty}^\infty (e^{g(s,x)} - 1)^2 \nu(dx)\right)\right] ds. \qquad (6.27)$$

6.2.1 Deterministic problem

We consider the case where the pair $(f_t, g(t,x))$ is non-random. Then from (6.27) we obtain

$$E[L_t^2] = 1 + \int_0^t E[L_s^2]\left(f_s^2 + \int_{-\infty}^\infty (e^{g(s,x)} - 1)^2 \nu(dx)\right) ds. \qquad (6.28)$$

Solving this equation, we obtain

$$E[L_T^2] = \exp\left(\int_0^T \left(f_s^2 + \int_{-\infty}^\infty (e^{g(s,x)} - 1)^2 \nu(dx)\right) ds\right). \qquad (6.29)$$

Now our problem is reduced to the following deterministic problem (DP).

Problem (DP): Find a constant f^* and a deterministic function $g^*(x)$ defined on $(-\infty, \infty)$ such that

$$(f^*)^2 + \int_{-\infty}^\infty (e^{g^*(x)} - 1)^2 \nu(dx) = \inf_{(f,g)\in\bar{\mathcal{C}}}\{f^2 + \int_{-\infty}^\infty (e^{g(x)} - 1)^2 \nu(dx)\}, \qquad (6.30)$$

where the set $\bar{\mathcal{C}}$ $(\subset \mathcal{C})$ consists of pairs of real number f and deterministic function $g(x)$ such that

$$b + \frac{1}{2}\sigma^2 + f\sigma + \int_{-\infty}^\infty (e^{g(x)}(e^x - 1) - x\mathbf{1}_{\{|x|\leq 1\}}(x))\nu(dx) = r. \qquad (6.31)$$

For the existence of a solution of this problem, we obtain the following theorem.

Theorem 6.1. (Sufficient Condition). *Assume that $(f^*, g^*(x), \gamma^*)$ satisfy the following conditions:*

$$f^* = \gamma^*\sigma \qquad (6.32)$$

$$e^{g^*(x)} - 1 = \gamma^*(e^x - 1) \qquad (6.33)$$

$$\gamma^*\sigma^2 + \int_{-\infty}^\infty \left((1 + \gamma^*(e^x - 1))(e^x - 1) - x\mathbf{1}_{\{|x|\leq 1\}}(x)\right)\nu(dx) = \beta, \qquad (6.34)$$

where

$$\beta = r - (b + \frac{1}{2}\sigma^2). \qquad (6.35)$$

Then $(f^*, g^*(x))$ is a unique solution of Problem (DP).

(Proof) Set the following:

$$k(x) = e^{g(x)} \qquad (k(x) > 0) \tag{6.36}$$

and

$$G(f,k) = f^2 + \int_{-\infty}^{\infty} \left\{ (k(x)-1)^2 \right\} \nu(dx). \tag{6.37}$$

Then the following condition (A) and assumption (B) are satisfied:

Condition (A):

$$f\sigma + \int_{-\infty}^{\infty} (k(x)(e^x - 1) - x 1_{\{|x| \le 1\}}(x))\nu(dx) = \beta, \tag{6.38}$$

$$f^*\sigma + \int_{-\infty}^{\infty} (k^*(x)(e^x - 1) - x 1_{\{|x| \le 1\}}(x))\nu(dx) = \beta. \tag{6.39}$$

Assumption (B):

$$f^* = \gamma^* \sigma, \tag{6.40}$$

$$(k^*(x) - 1) = \gamma^*(e^x - 1). \tag{6.41}$$

What we have to prove is

$$G(f,k) \ge G(f^*, k^*). \tag{6.42}$$

We remark here that $G(f,k)$ is convex w.r.t. (f,k). In fact, f^2 is a convex function of f and $(k-1)^2$ is a convex function of k. So we obtain

$$G(f,k) - G(f^*, k^*)$$

$$\ge 2f^*(f - f^*) + 2\int_{-\infty}^{\infty} (k^*(x) - 1))\,(k(x) - k^*(x))\nu(dx)$$

$$= 2\gamma^*\sigma(f - f^*) + 2\int_{-\infty}^{\infty} \{\gamma^*(e^x - 1)\}\,(k(x) - k^*(x))\nu(dx), \tag{6.43}$$

where we have used the convexity of $G(f,k)$ for the inequality and assumptions (6.40) and (6.41) for the equality. The value of the last formula is equal to 0 by (6.38) and (6.39).

Since (6.42) is proved, (f^*, g^*) is a solution of Problem (DP). The uniqueness of the solution follows from the strong convexity of $G(f,k)$. (Q.E.D.)

Remark 6.2. If a solution exists, then from (6.33) in Theorem 6.1, $\gamma^*(e^x - 1) + 1$ must be in $(0, \infty)$. In the case where the support of ν is $\Re$, this condition is equivalent to $0 \le \gamma^* < 1$, or in terms of the generating triplet of the process Z_t, to

$$0 \le \beta - \int_{-\infty}^{\infty} ((e^x - 1) - x 1_{\{|x| \le 1\}}(x))\nu(dx) < \sigma^2 + \int_{-\infty}^{\infty} (e^x - 1)^2 \nu(dx). \tag{6.44}$$

Next we investigate necessary conditions for the existence of a solution of Problem (DP). In the case of $\sigma \neq 0$, we obtain the following theorem.

Theorem 6.2. *Assume that $\sigma \neq 0$ and Problem (DP) has a solution $(f^*, g^*(x))$. Then $(f^*, g^*(x))$ is given by*

$$f^* = \gamma^* \sigma \tag{6.45}$$

$$e^{g^*(x)} - 1 = \gamma^*(e^x - 1), \tag{6.46}$$

where γ^* is the solution of the following equation:

$$\gamma \sigma^2 + \int_{-\infty}^{\infty} \left((1 + \gamma(e^x - 1))(e^x - 1) - x 1_{\{|x| \leq 1\}}(x) \right) \nu(dx) = \beta. \tag{6.47}$$

(Proof) Let us introduce a function $k(x) = e^{g(x)}$. Then the condition that $(f, g) \in \bar{\mathcal{C}}$ is equivalent to the following condition:

$$f\sigma + \int_{-\infty}^{\infty} (k(x)(e^x - 1) - x 1_{\{|x| \leq 1\}}(x)) \nu(dx) = \beta, \text{ and } k(x) > 0. \tag{6.48}$$

We denote by $\widetilde{\mathcal{C}}$ the set of all (f, k), which satisfies (6.48). Since this condition is linear in f and k, the set $\widetilde{\mathcal{C}}$ is convex.

Set $k^*(x) = e^{g^*(x)}$, then $(f^*, k^*) \in \widetilde{\mathcal{C}}$ follows from $(f^*, g^*) \in \bar{\mathcal{C}}$. Take a real number ϕ and a function $\psi(x)$ such that $|\psi(x)| < a k^*(x)$ for some $a > 0$ and satisfies the following condition:

$$\phi\sigma + \int_{-\infty}^{\infty} \psi(x)(e^x - 1)\nu(dx) = 0. \tag{6.49}$$

Then it follows that

$$(f^* + \epsilon\phi, k^* + \epsilon\psi(x)) \in \widetilde{\mathcal{C}} \quad if \ |\epsilon| \ is \ small. \tag{6.50}$$

We fix such a pair $(\phi, \psi(x))$, and set

$$H_{\phi,\psi}(\epsilon) = G(f^* + \epsilon\phi, k^*(x) + \epsilon\psi(x))$$
$$= (f^* + \epsilon\phi)^2 + \int_{-\infty}^{\infty} (k^*(x) + \epsilon\psi(x) - 1)^2 \nu(dx). \tag{6.51}$$

Since (f^*, k^*) is a minimal point of $G(f, k)$, it holds that $\frac{dH_{\phi,\psi}}{d\epsilon}(0) = 0$. From the following formula:

$$\frac{dH_{\phi,\psi}}{d\epsilon}(\epsilon) = 2(f^* + \epsilon\phi)\phi + \int_{-\infty}^{\infty} 2(k^*(x) + \epsilon\psi(x) - 1)\psi\nu(dx), \tag{6.52}$$

we obtain the following equality:

$$f^*\phi + \int_{-\infty}^{\infty} (k^*(x) - 1)\psi(x)\nu(dx) = 0. \tag{6.53}$$

We remark here that $\psi(x)$ can be taken freely under the condition $|\psi(x)| < ak^*(x)$ for some $a > 0$, because for such a function $\psi(x)$ the condition (6.49) is satisfied by setting the constant $\phi = -\frac{1}{\sigma} \int_{-\infty}^{\infty} \psi(x)(e^x - 1)\nu(dx)$.

From (6.53), we obtain

$$\int_{-\infty}^{\infty} (f^*(e^x - 1) - \sigma(k^*(x) - 1))\psi(x)\nu(dx) = 0. \tag{6.54}$$

Since $\psi(x)$ can be taken freely under the condition that $|\psi(x)| < ak^*(x)$ for some $a > 0$, the following equality is obtained:

$$f^*(e^x - 1) - \sigma(k^*(x) - 1) = 0, \quad \nu - a.e. \tag{6.55}$$

In fact, if $f^*(e^x - 1) - \sigma(k^*(x) - 1) > 0$ on a set A^+ such that $\nu(A^+) > 0$, set $\tilde{\psi}(x) = k^*(x)1_{A^+}(x)$. Then $|\tilde{\psi}(x)| \leq k^*(x)$ and

$$\int_{-\infty}^{\infty} (f^*(e^x - 1) - \sigma(k^*(x) - 1))\psi(x)\nu(dx) > 0. \tag{6.56}$$

This contradicts (6.54).

Next we set $\gamma^* = f^*/\sigma$, then conditions (6.32) and (6.33) in Theorems 6.1. Using (6.55), we know that γ^* is a solution of (6.34). (Q.E.D.)

Remark 6.3. Combining Theorem 6.1 and 6.2, we get a necessary and sufficient condition for the existence of a solution of Problem (DP) in the case of $\sigma \neq 0$.

In the case of $\sigma = 0$, we need some additional assumptions.

Suppose that Problem (DP) has a solution $g^*(x)$, and take a function $\psi(x) \in L^2(\nu(dx))$ such that $|\psi(x)| < ak^*(x)$ for some $a > 0$ and

$$\int_{-\infty}^{\infty} \psi(x)(e^x - 1)\nu(dx) = 0. \tag{6.57}$$

Then, as the same way as we have seen above in the proof of Theorem 6.2 for the minimal point k^*, it holds that

$$\int_{-\infty}^{\infty} (k^*(x) - 1)\psi(x)\nu(dx) = 0. \tag{6.58}$$

Set the following:

$$\mathcal{L}_0 = \left\{ \alpha(e^x - 1) \in L^2(\nu(dx)); -\infty < \alpha < \infty \right\}, \tag{6.59}$$

then $\mathcal{L}_0$ is a one-dimensional linear subspace of $L^2(\nu(dx))$.

Next we set

$$\mathcal{L}_1 = \left\{ \psi(x) \in L^2(\nu(dx)); \psi(x) \ satisfies \ (6.57) \right.$$

$$\left. and \ |\psi(x)| < ak^*(x) \ for \ some \ a > 0 \right\}. \qquad (6.60)$$

$$\mathcal{L}_2 = \{ \psi(x) \in L^2(\nu(dx)); \psi(x) \ satisfies \ (6.57) \}. \qquad (6.61)$$

Then $\mathcal{L}_1$ and $\mathcal{L}_2$ are linear subspaces of $L^2(\nu(dx))$ and $\mathcal{L}_1 \subseteq \mathcal{L}_2$. Here we need an assumption.

Assumption (C): For any $k(x) \in L^2(\nu(dx))$ such that $k(x) > 0$, the set $\mathcal{L}_1(k)$ defined by (6.60) is dense in $\mathcal{L}_2$.

Theorem 6.3. *Assume that* $\sigma = 0$ *and* Assumption (C) *is satisfied. In this case, if Problem (DP) has a solution* $g^*(x)$, *then* $g^*(x)$ *is given by*

$$e^{g^*(x)} - 1 = \gamma^*(e^x - 1), \qquad (6.62)$$

where γ^* *is the solution of the following equation:*

$$\int_{-\infty}^{\infty} \left((1 + \gamma(e^x - 1)) (e^x - 1) - x 1_{\{|x| \le 1\}}(x) \right) \nu(dx) = \beta. \qquad (6.63)$$

(Proof) Let $g^*(x))$ be a solution, and set $k^*(x) = e^{g^*(x)}$. The condition (6.57) means that $\psi(x)$ is orthogonal to $(e^x - 1)$, and the condition (6.58) means that $(k^*(x) - 1)$ is orthogonal to $\psi(x)$. Using Assumption (C) we know that $\mathcal{L}_0^{\perp} = \mathcal{L}_2$ and $(k^*(x) - 1) \in \mathcal{L}_2^{\perp}$. Since $\mathcal{L}_2^{\perp} = (\mathcal{L}_0^{\perp})^{\perp} = \mathcal{L}_0$, so we obtain $(k^*(x) - 1) = \gamma^*(e^x - 1)$ for some constant γ^*. The equation (6.63) follows from (6.34) with $\sigma = 0$. (Q.E.D.)

Remark 6.4. From Theorems 6.2 and 6.3, we know that the sufficient condition in Theorem 6.1 is almost necessary for the existence of a solution of Problem (DP).

6.2.2   *Existence theorem of the MVEMM*

Our original problem is finding a predictable process f_t^* and a predictable function process $g_t^*(x)$ such that $(f^*, g^*) \in \mathcal{C}$ and

$$E[(L_T(f^*, g^*))^2] = \inf_{(f,g) \in \mathcal{C}} E[(L_T(f, g))^2]. \qquad (6.64)$$

We see a relation between $\bar{\mathcal{C}}$ and $\mathcal{C}$. Define

$$H(f, g) = f^2 + \int_{-\infty}^{\infty} (e^{g(x)} - 1)^2 \nu(dx) \quad for \quad (f, g) \in \bar{\mathcal{C}}, \qquad (6.65)$$

and set

$$H_0 = \inf\{H(f,g); \quad (f,g) \in \bar{C}\}. \tag{6.66}$$

Then it easy to see that

$$f_s^2 + \int_{-\infty}^{\infty} (e^{g(s,x)} - 1)^2 \nu(dx) \geq H_0 \quad P(d\omega)ds - a.s., \quad \forall (f,g) \in C. \tag{6.67}$$

Using this fact, we obtain the following lemma.

Lemma 6.1. (*i*) *Under the above notations it holds that*

$$E[L_T^2(f,g)] \geq e^{H_0 T}, \quad \forall (f,g) \in C. \tag{6.68}$$

(*ii*) *For any* $(f,g) \in C$, *there exists a* $(\bar{f}, \bar{g}) \in \bar{C}$ *such that*

$$E[L_T^2(\bar{f}, \bar{g})] \leq E[L_T^2(f,g)] \tag{6.69}$$

(Proof) (i) From (6.67) it follows that

$$E[L_t^2] = 1 + \int_0^t E\left[L_s^2 \left(f_s^2 + \int_{-\infty}^{\infty} (e^{g(s,x)} - 1)^2 \nu(dx)\right)\right] ds$$

$$\geq 1 + \int_0^t E[L_s^2] H_0 ds. \tag{6.70}$$

The result (i) follows from Gronwall's lemma.

(ii) **Case 1.** If there exists a $(f^*, g^*) \in \bar{C}$ such that $H(f^*, g^*) = H_0$, then from (i) we obtain

$$E[L_T^2(f^*, g^*)] = e^{H_0 T} \leq E[L_T^2(f,g)]. \tag{6.71}$$

So, taking $\bar{f} = f^*$ and $\bar{g} = g^*$, (6.69) is proved.

Case 2. Assume that the value H_0 is not attained by any element in $\bar{C}$. In such a case, for any $\epsilon > 0$ we can find an element $(f^{(\epsilon)}, g^{(\epsilon)}) \in \bar{C}$ such that

$$H(f^{(\epsilon)}, g^{(\epsilon)}) \leq H_0 + \epsilon. \tag{6.72}$$

From (6.72) it follows that

$$E\left[L_t^2(f^{(\epsilon)}, g^{(\epsilon)})\right] = 1 + \int_0^t E\left[L_s^2(f^{(\epsilon)}, g^{(\epsilon)}) H(f^{(\epsilon)}, g^{(\epsilon)})\right] ds$$

$$\leq 1 + \int_0^t E\left[L_s^2(f^{(\epsilon)}, g^{(\epsilon)})\right] (H_0 + \epsilon) ds. \tag{6.73}$$

Using Gronwall's lemma we get

$$E[L_T^2(f^{(\epsilon)}, g^{(\epsilon)})] \leq e^{(H_0 + \epsilon)T}. \tag{6.74}$$

From the assumption that the value H_0 is not attained, the following inequality holds:

$$H(f_s(\omega), g_s(\cdot, \omega)) > H_0 \quad P(d\omega)ds - a.s. \tag{6.75}$$

Set the following:

$$\Omega_{(\delta,t)} = \{\omega \in \Omega; H(f_t(\omega), g_t(\cdot, \omega)) > H_0 + \delta\}, \tag{6.76}$$

then it holds that

$$\lim_{\delta \downarrow 0} \Omega_{(\delta,t)} = \Omega \quad (P \, a.s.) \, for \, a.a. \, t. \tag{6.77}$$

Using the facts noticed above, the following calculation is easy:

$$\begin{aligned}
E[L_t^2(f,g)] &= 1 + \int_0^t E[L_s^2(f,g)H(f_s(\omega), g_s(\cdot, \omega))]ds \\
&= 1 + \int_0^t E[L_s^2(f,g)H(f_s(\omega), g_s(\cdot, \omega))1_{\Omega_{(\delta,s)}}]ds \\
&\quad + \int_0^t E[L_s^2(f,g)H(f_s(\omega), g_s(\cdot, \omega))1_{\Omega_{(\delta,s)}^C}]ds \\
&\geq 1 + \int_0^t E[L_s^2(f,g)1_{\Omega_{(\delta,s)}}](H_0 + \delta)ds \\
&\quad + \int_0^t E[L_s^2(f,g)1_{\Omega_{(\delta,s)}^C}]H_0 ds \\
&= 1 + \int_0^t E[L_s^2(f,g)]H_0 ds + \delta \int_0^t E[L_s^2(f,g)1_{\Omega_{(\delta,s)}}]ds.
\end{aligned} \tag{6.78}$$

Set the following:

$$\chi_\delta(t) = \delta \int_0^t E[L_s^2(f,g)1_{\Omega_{(\delta,s)}}]ds, \tag{6.79}$$

then, taking δ small enough, we obtain from (6.77) that

$$\chi_\delta(t) = \int_0^t E[L_s^2(f,g)1_{\Omega_{(\delta,s)}}]ds > 0. \tag{6.80}$$

So we get the following inequality:

$$E[L_t^2(f,g)] \geq 1 + \chi_\delta(t) + \int_0^t E[L_s^2(f,g)]H_0 ds. \tag{6.81}$$

Using the extended Gronwall's lemma and the fact that $\chi(s) \geq 0$, we obtain

$$E[L_t^2(f,g)] \geq (1 + \chi_\delta(t)) + \int_0^t H_0(1 + \chi(s))e^{\int_s^t H_0 du} ds$$

$$\geq (1 + \chi_\delta(t)) + H_0 \int_0^t e^{H_0(t-s)} ds$$

$$= e^{H_0 t} + \chi_\delta(t). \tag{6.82}$$

Thus we have finally obtained

$$E[L_T^2(f,g)] > e^{H_0 T}. \tag{6.83}$$

Comparing (6.74) and (6.83), we can choose a positive value for ϵ small enough so that the following inequality holds:

$$E[L_T^2(f^{(\epsilon)}, g^{(\epsilon)})] < E[L_T^2(f,g)]. \tag{6.84}$$

Taking $\bar{f} = f^{(\epsilon)}$ and $\bar{g} = g^{(\epsilon)}$ the proof of (ii) is completed. (Q.E.D.)

Remark 6.5. It holds that

$$E[L_t^2(f,g)] = e^{H(f,g)t} \quad \forall (f,g) \in \bar{C}. \tag{6.85}$$

This lemma means that the optimization problem in C is reduced to the optimization problem in $\bar{C}$, and we obtain the following theorem.

Theorem 6.4. (i) *For the existence of the MVEMM it is necessary and sufficient that Problem (DP) has a solution.*
(ii) *Assume that Problem (DP) has a solution* $(\bar{f}^*, \bar{g}^*)$, *then* $Q^{L_T(\bar{f}^*, \bar{g}^*)}$ *is the MVEMM for* S_t.

(Proof) (i) Assume the existence of the MVEMM and let it be given by $Q^{L_T(f^*, g^*)}, (f^*, g^*) \in C$. From Lemma 6.1 (ii) it follows that there exists $(\bar{f}^*, \bar{g}^*) \in \bar{C}$ such that

$$E[L_T^2(\bar{f}^*, \bar{g}^*)] \leq E[L_T^2(f^*, g^*)] = \inf_{(f,g) \in C}\{E[L_T^2(f,g)]\}. \tag{6.86}$$

This formula proves that $Q^{(\bar{f}^*, \bar{g}^*)}$ is the MVEMM and that (f^*, g^*) is the minimal point in C. Since $(\bar{f}^*, \bar{g}^*) \in \bar{C}$ and $\bar{C} \subset C$, $(\bar{f}^*, \bar{g}^*)$ is the solution of (DP).

Conversely, assume that Problem (DP) has a solution $(\bar{f}^*, \bar{g}^*) \in \bar{C}$. Then, by Remark 6.5 and Lemma 6.1 (i), we obtain

$$E[L_T^2(\bar{f}^*, \bar{g}^*)] = e^{H_0 T} \leq E[L_T^2(f,g)] \quad \forall (f,g) \in C. \tag{6.87}$$

Therefore $Q^{L_T(\bar{f}^*,\bar{g}^*)}$ is the MVEMM.

(ii) This result is already proved in the proof of (i).

(Q.E.D.)

Discussions above lead to the following theorem:

Theorem 6.5. *Assume that there exists a function $\bar{g}^*(x)$ which satisfies the following condition:*

$$e^{\bar{g}^*(x)} - 1 = \gamma^*(e^x - 1), \tag{6.88}$$

where γ^* is a solution of

$$\gamma^* \sigma^2 + \int_{-\infty}^{\infty} \left((1 + \gamma^*(e^x - 1))(e^x - 1) - x1_{\{|x|\leq 1\}}(x) \right) \nu(dx) = \beta. \tag{6.89}$$

Then, setting $\bar{f}^* = \gamma^* \sigma$, $Q^{L_T(\bar{f}^*,\bar{g}^*)}$ is the MVEMM for S_t.

(Proof) By Theorem 6.1, $(\bar{f}^*, \bar{g}^*)$ is a solution of Problem (DP). Therefore from Theorem 6.4 (ii), $Q^{L_T(\bar{f}^*,\bar{g}^*)}$ is the MVEMM for S_t.

(Q.E.D.)

Remark 6.6. Theorem 6.5 means that the existence of $\bar{g}^*(x)$ and γ^*, which satisfy the conditions of (6.88) and (6.89), is a sufficient condition for the existence of the MVEMM. On the other hand, from the results of Theorem 6.2, 6.3, and Remark 6.4, we can say that the existence of $\bar{g}^*(x)$ and γ^*, which satisfy the conditions of (6.88) and (6.89), is almost necessary for the existence of the MVEMM.

6.2.3 Generating triplet of Z_t under MVEMM

Assume that the assumptions of Theorem 6.1 are satisfied, and let $(f^*, g^*(x))$ be the solution of (DP). Then $L_t^* = L_t(f^*, g^*)$ is a solution of the following equation:

$$dL_t^* = L_{t-}^* \left(f^* dW_t + \int_{-\infty}^{\infty} (e^{g^*(x)} - 1)\, \widetilde{N}(dt, dx) \right)$$

$$= L_{t-}^* \left(\gamma^* \sigma dW_t + \int_{-\infty}^{\infty} \gamma^*(e^x - 1)\, \widetilde{N}(dt, dx) \right), \tag{6.90}$$

where

$$\gamma^* = \frac{\beta - \int_{-\infty}^{\infty}(e^x - 1 - x1_{\{|x|\leq 1\}}(x))\nu(dx)}{\sigma^2 + \int_{-\infty}^{\infty}(e^x - 1)^2 \nu(dx)}. \tag{6.91}$$

By (6.5) the explicit form of L_t^* is

$$L_t^* = \exp\left\{ f^*W_t - \left(\frac{1}{2}(f^*)^2 + \int_{-\infty}^{\infty} (e^{g^*(x)} - 1 - g^*(x))\nu(dx)\right) t \right.$$
$$\left. + \int^t \int_{-\infty}^{\infty} g^*(x) \, \widetilde{N}(ds, dx) \right\}. \tag{6.92}$$

The Lévy measure of Z_t under $Q^{L_T(f^*, g^*)}$ is, from (6.13),

$$e^{g^*(x)}\nu(dx) = (1 + \gamma^*(e^x - 1))\,\nu(dx). \tag{6.93}$$

Thus we have obtained the following theorem.

Theorem 6.6. *Assume that* $(f^*, g^*(x), \gamma^*)$ *is a solution of the following equations:*

$$f^* = \gamma^*\sigma, \tag{6.94}$$
$$e^{g^*(x)} - 1 = \gamma^*(e^x - 1), \tag{6.95}$$
$$\gamma^*\sigma^2 + \int_{-\infty}^{\infty} \left((1 + \gamma^*(e^x - 1))(e^x - 1) - x1_{|x|\leq 1}\right)\nu(dx) = \beta. \tag{6.96}$$

Then
(i) $Q^{L_T(f^*, g^*)}$ *is the MVEMM for* S_t,
(ii) the Lévy measure $\nu^{(MVEMM)}(dx)$ *of* Z_t *under* $P^{(MVEMM)}$ *is*

$$\nu^{(MVEMM)}(dx) = e^{g^*(x)}\nu(dx) = (1 + \gamma^*(e^x - 1))\,\nu(dx). \tag{6.97}$$

(Proof) By Theorem 6.1 $(f^*, g^*(x), \gamma^*)$ is a solution of Problem (DP). Therefore by Theorem 6.4 $Q^{L_T(f^*, g^*)}$ is the MVEMM for S_t. The form of the Lévy measure $\nu^{(MVEMM)}(dx)$ is already obtained in (6.93). (Q.E.D.)

Remark 6.7. The generating triplet of Z_t under $P^{(MVEMM)}$ is determined in the following way: $\sigma^{(MVEMM)}$ is not changed $(\sigma^{(MVEMM)} = \sigma)$, $\nu^{(MVEMM)}(dx)$ is given by 6.6 (ii), and $b^{(MVEMM)}$ is determined from the martingale condition.

6.3   The Minimal L^q Equivalent Martingale Measure

In this section we investigate the minimal L^q equivalent martingale measure (ML^{q}EMM) for $q \in (-\infty, \infty), q \neq 0, 1$. (The case of $q = 2$ is already studied in the previous section, and ML^{2}EMM is identified with MVEMM.)

6.3.1   *The case of $q > 1$*

The distance function of this case is $F(x) = x^q$, which corresponds to a utility function $u(x) = \frac{-1}{p}x^p$, $p > 1$ as explained in Section 5.3. (See Remark 5.2.) The conjugate function is $u^*(y) = \frac{1}{q}y^q$, $q = \frac{p}{p-1} > 1$. The equation (6.24) for $L_t = L_t(f, g)$ is, corresponding to $F(x) = x^q$, in the following form:

$$
E[L_t^q] = 1 + \int_0^t E\left[L_s^q\left(\frac{1}{2}q(q-1)f_s^2\right.\right.
$$
$$
\left.\left. + \int_{-\infty}^{\infty}\left\{e^{qg(s,x)} - 1 - q(e^{g(s,x)} - 1)\right\}\nu(dx)\right)\right]ds. \quad (6.98)
$$

If the pair $(f_t, g(t, x))$ is non-random, then we obtain

$$
E[L_t^q] = 1 + \int_0^t E[L_s^q]\left(\frac{1}{2}q(q-1)f_s^2\right.
$$
$$
\left. + \int_{-\infty}^{\infty}\left\{e^{qg(s,x)} - 1 - q(e^{g(s,x)} - 1)\right\}\nu(dx)\right)ds. \quad (6.99)
$$

Solving this equation for $E[L_t^q]$, we obtain

$$
E[L_T^q] = \exp\left\{\int_0^T\left(\frac{1}{2}q(q-1)f_s^2\right.\right.
$$
$$
\left.\left. + \int_{-\infty}^{\infty}\left\{e^{qg(s,x)} - 1 - q(e^{g(s,x)} - 1)\right\}\nu(dx)\right)ds\right\}. \quad (6.100)
$$

So the corresponding deterministic problem is the following problem $(\mathrm{DP})_1^{(q)}$, $q > 1$.

Problem $(\mathrm{DP})_1^{(q)}$, $q > 1$: Find a constant f^* and a deterministic function $g^*(x)$ defined on $(-\infty, \infty)$ such that

$$
\frac{1}{2}q(q-1)(f^*)^2 + \int_{-\infty}^{\infty}\left\{e^{qg^*(x)} - 1 - q(e^{g^*(x)} - 1)\right\}\nu(dx)
$$
$$
= \inf\left\{\frac{1}{2}q(q-1)f^2 + \int_{-\infty}^{\infty}\left\{e^{qg(x)} - 1 - q(e^{g(x)} - 1)\right\}\nu(dx), (f,g) \in \bar{\mathcal{C}}\right\}. \quad (6.101)
$$

We can carry out a similar analysis on this problem as was done to the MVEMM in the previous section, and we obtain the following theorems as a result.

Theorem 6.7. (i) *For the existence of the $ML^q EMM$, $q > 1$, it is necessary and sufficient that Problem $(DP)_1^{(q)}$, $q > 1$, has a solution.*
(ii) *When the condition of* (i) *is satisfied, let $(f^*, g^*(x))$ be the deterministic solution. Then $Q^{L_T(f^*, g^*)}$ is the $ML^q EMM$ for S_t.*

Theorem 6.8. *Assume that $(f_q^*, g_q^*(x))$ satisfies the following conditions:*

$$q(q-1)f_q^* = \mu_q^* \sigma \qquad (6.102)$$

$$q(e^{(q-1)g_q^*(x)} - 1) = \mu_q^*(e^x - 1) \qquad (6.103)$$

where μ_q^* is the solution of

$$\frac{\mu_q \sigma^2}{q(q-1)} + \int_{-\infty}^{\infty} \left((1 + \frac{\mu_q(e^x - 1)}{q})^{\frac{1}{q-1}} (e^x - 1) - x 1_{\{|x| \leq 1\}}(x) \right) \nu(dx) = \beta. \qquad (6.104)$$

Then the martingale measure $Q^{(f_q^*, g_q^*)}$ is the $ML^q EMM$, we denote it by $P^{(ML^q EMM)}$, and the Lévy measure of Z_t under $P^{(ML^q EMM)}$ is

$$\nu^{(ML^q EMM)}(dx) = e^{g_q^*(x)} \nu(dx) = \left(1 + \frac{\mu_q^*}{q}(e^x - 1) \right)^{\frac{1}{q-1}} \nu(dx). \qquad (6.105)$$

Remark 6.8. Set $\gamma_q^* = \frac{\mu_q^*}{q(q-1)}$, then γ_q^* is a solution of the following equation:

$$\gamma \sigma^2 + \int_{-\infty}^{\infty} \left((1 + (q-1)\gamma(e^x - 1))^{\frac{1}{q-1}} (e^x - 1) - x 1_{\{|x| \leq 1\}}(x) \right) \nu(dx) = \beta. \qquad (6.106)$$

And (6.102), (6.103) and (6.105) in Theorem 6.8 are expressed in the following forms:

$$f_q^* = \gamma_q^* \sigma, \qquad (6.107)$$

$$e^{g_q^*(x)} = \left(1 + (q-1)\gamma_q^*(e^x - 1) \right)^{\frac{1}{q-1}} \qquad (6.108)$$

$$\nu^{(ML^q EMM)}(dx) = \left(1 + (q-1)\gamma_q^*(e^x - 1) \right)^{\frac{1}{q-1}} \nu(dx). \qquad (6.109)$$

These forms are convenient when we compare the $ML^q EMM$ with the MEMM (see Theorem 6.12).

6.3.2 The case of $0 < q < 1$

In this case the corresponding utility function is $u(x) = \frac{1}{p}x^p$, $p < 0$, and the conjugate function is $u^*(y) = \frac{-1}{q}y^q$, $0 < q = \frac{p}{p-1} < 1$. (See section 5.3.) The distance function is $F(x) = -x^q, x > 0, 0 < q < 1$. Since this function is convex, we can apply a similar analysis in this case as we have done in Section 6.2.

Remark 6.9. In this case the ML^qEMM, denoted by $P^{(ML^qEMM)}$, is not minimal in the sense of $\inf E_P[L_T^q(f,g)]$. It is minimal in the sense of $\inf\{E_P[-L_T^q(f,g)]\}$.

From this remark, we know that the corresponding deterministic problem is the following $(\mathrm{DP})_2^{(q)}$, $0 < q < 1$.

Problem $(\mathrm{DP})_2^{(q)}$, $0 < q < 1$: Find a constant f^* and a deterministic function $g^*(x)$ defined on $(-\infty, \infty)$ such that

$$\frac{1}{2}q(q-1)(f^*)^2 + \int_{-\infty}^{\infty} \left\{ e^{qg^*(x)} - 1 - q(e^{g^*(x)} - 1) \right\} \nu(dx)$$

$$= \sup \left\{ \frac{1}{2}q(q-1)f^2 + \int_{-\infty}^{\infty} \left\{ e^{qg(x)} - 1 - q(e^{g(x)} - 1) \right\} \nu(dx), (f,g) \in \bar{\mathcal{C}} \right\}.$$
$$(6.110)$$

We can do a similar analysis on this problem as we have done to the MVEMM in the previous section (or in the previous subsections), and obtain the following theorem.

Theorem 6.9. (i) *For the existence of the ML^qEMM, $0 < q < 1$, it is necessary and sufficient that Problem $(DP)_2^{(q)}$, $0 < q < 1$, has a solution.*
(ii) *When the condition of (i) is satisfied, let (f^*, g^*) be the deterministic solution. Then $Q^{L_T(f^*,g^*)}$ is the ML^qEMM for S_t.*
(iii) *Theorem 6.8 holds true in this case too.*

6.3.3 *The case of $q < 0$*

In this case the corresponding utility function is $u(x) = \frac{1}{p}x^p$, $0 < p < 1$, and the conjugate function is $u^*(y) = \frac{-1}{q}y^q$, $q = \frac{p}{p-1} < 0$. (See Section 5.3.) So the distance function is $F(x) = x^q$, $x > 0$, $q < 0$, and the corresponding deterministic problem is the following $(\mathrm{DP})_3^{(q)}$, $q < 0$.

Problem $(\mathrm{DP})_3^{(q)}$, $q < 0$: Find a constant f^* and a deterministic function g^* defined on $(-\infty, \infty)$ such that

$$\frac{1}{2}q(q-1)(f^*)^2 + \int_{-\infty}^{\infty} \left\{ e^{qg^*(x)} - 1 - q(e^{g^*(x)} - 1) \right\} \nu(dx)$$

$$= \inf \left\{ \frac{1}{2}q(q-1)f^2 + \int_{-\infty}^{\infty} \left\{ e^{qg(x)} - 1 - q(e^{g(x)} - 1) \right\} \nu(dx), (f,g) \in \bar{\mathcal{C}} \right\}.$$

(6.111)

We can do a similar analysis on this problem as we have done to the MVEMM in the previous section (or in the previous subsection), and obtain the following theorem.

Theorem 6.10. (i) *For the existence of the $ML^q EMM$, $q < 0$, it is necessary and sufficient that Problem $(DP)_3^{(q)}$, $q < 0$, has a solution.*
(ii) *When the condition of (i) is satisfied, let (f^*, g^*) be the deterministic solution. Then $Q^{L_T(f^*,g^*)}$ is the $ML^q EMM$ for S_t.*
(iii) *Theorem 6.8 holds true in this case too.*

6.4 Minimal Entropy Martingale Measures

Let us investigate the minimal entropy martingale measure (MEMM). The distance function $F(x)$ for the MEMM is

$$F(x) = x \log x,$$

(6.112)

and from the formula (6.24) we obtain

$$E[F(L_t)] = \frac{1}{2} \int_0^t E\left[L_{s-} f_s^2 \right] ds$$
$$+ \int_0^t \left(\int_{-\infty}^{\infty} E\left[L_{s-} \left(g_s(x)e^{g_s(x)} - e^{g_s(x)} + 1 \right) \right] \nu(dx) \right) ds.$$

(6.113)

We consider the case where f_s and $g_s(x)$ are deterministic. Then we obtain the following formula:

$$E[F(L_t)] = \int_0^t E\left[L_{s-} \right] \left(\frac{1}{2}f_s^2 + \int_{-\infty}^{\infty} \left(g_s(x)e^{g_s(x)} - e^{g_s(x)} + 1 \right) \nu(dx) \right) ds$$
$$= \int_0^t \left(\frac{1}{2}f_s^2 + \int_{-\infty}^{\infty} \left(g_s(x)e^{g_s(x)} - e^{g_s(x)} + 1 \right) \nu(dx) \right) ds, \quad (6.114)$$

and we get to the following deterministic problem $(DP)^{(entropy)}$.

Problem $(\mathbf{DP})^{(entropy)}$: Find a constant f^* and a deterministic function $g^*(x)$ defined on $(-\infty, \infty)$ such that

$$\frac{1}{2}(f^*)^2 + \int_{-\infty}^{\infty} \left(e^{g^*(x)}(g^*(x) - 1) + 1 \right) \nu(dx)$$

$$= \inf \left\{ \frac{1}{2}f^2 + \int_{-\infty}^{\infty} \left(e^{g(x)}(g(x) - 1) + 1 \right) \nu(dx), \ (f,g) \in \bar{C} \right\}. \quad (6.115)$$

We can do a similar analysis to this problem as we have done to the MVEMM or ML^{q}EMM in the previous sections, and obtain the following theorems.

Theorem 6.11. (i) *For the existence of the MEMM it is necessary and sufficient that Problem* $(DP)^{(entropy)}$ *has a solution.*
(ii) *Assume that the condition of* (i) *is satisfied, and let* (f^*, g^*) *be the deterministic solution. Then* $Q^{L_T(f^*,g^*)}$ *is the MEMM for* S_t.

Theorem 6.12. *Assume that* $(f^*, g^*(x))$ *satisfies the following conditions:*

$$f^* = \gamma^* \sigma, \quad (6.116)$$

$$g^*(x) = \gamma^*(e^x - 1), \quad (6.117)$$

where γ^* *is the solution of*

$$\gamma \sigma^2 + \int_{-\infty}^{\infty} \left(e^{\gamma(e^x - 1)}(e^x - 1) - x1_{\{|x| \le 1\}}(x) \right) \nu(dx) = \beta. \quad (6.118)$$

Then the martingale measure $Q^{(f^*,g^*)}$ *is the MEMM, which is denoted by* $P^{(MEMM)}$, *and the Lévy measure of* Z_t *under* $P^{(MEMM)}$ *is*

$$\nu^{(MEMM)}(dx) = e^{g^*(x)}\nu(dx) = e^{\gamma^*(e^x - 1)}\nu(dx). \quad (6.119)$$

The minimal entropy is

$$H(Q^{(f^*,g^*)}|P) = E[F(L_T(f^*, g^*)]$$

$$= T \left\{ \frac{1}{2}\sigma^2 \gamma^{*2} \int_{-\infty}^{\infty} \left(\gamma^*(e^x - 1)e^{\gamma^*(e^x - 1)} - e^{\gamma^*(e^x - 1)} + 1 \right) \nu(dx) \right\}.$$

$$(6.120)$$

Since $\beta = r - (b + \frac{1}{2}\sigma^2)$, the equation (6.118) is just the same as the equation (4.28) in Theorem 4.2. So the solution γ^* is identified with the solution $\hat{h}^*$, and we obtain

$$\frac{dP^{(MEMM)}}{dP}\Big|_{\mathcal{F}_T} = L_T(f^*, g^*)$$

$$= \exp\left\{\gamma^*\sigma W_t - \frac{1}{2}\gamma^{*2}\sigma^2 t + \int_0^t \int_{-\infty}^{\infty} \gamma^*(e^x - 1)\tilde{N}_p(dsdx) \right.$$

$$\left. - \left(\int_{-\infty}^{\infty}\left(e^{\gamma^*(e^x-1)} - 1 - \gamma^*(e^x - 1)\right)\nu(dx)\right)t\right\}$$

$$= \exp\left\{\hat{h}^*\sigma W_t - \frac{1}{2}\hat{h}^{*2}\sigma^2 t + \int_0^t \int_{-\infty}^{\infty} \hat{h}^*(e^x - 1)\tilde{N}_p(dsdx) \right.$$

$$\left. - \left(\int_{-\infty}^{\infty}\left(e^{\hat{h}^*(e^x-1)} - 1 - \hat{h}^*(e^x - 1)\right)\nu(dx)\right)t\right\}$$

$$= \frac{d\hat{P}^{(ESSMM)}}{dP}\Big|_{\mathcal{F}_T}. \qquad (6.121)$$

Thus we have obtained the following result.

Theorem 6.13. *The simple-return Esscher-transformed martingale measure* $P_{\hat{Z}_{[0,T]}}^{(ESSMM)} = \hat{P}^{(ESSMM)}$ *of* S_t *is identified with the minimal entropy martingale measure* $P^{(MEMM)}$ *of* S_t.

Remark 6.10. Based on the result that $\hat{P}^{(ESSMM)} = P^{(MEMM)}$, the "simple-return Esscher martingale measure" is usually called the "minimal entropy martingale measure" (MEMM), and we denote this measure by $P^{(MEMM)}$ or sometimes by P^* for simplicity. (See Sections 4.3 and 4.4.)

6.5   Convergence of ML^{q}EMM to MEMM (as $q \downarrow 1$)

Assume that ML^{q}EMM exists for $1 < q \le 2$ and that the assumptions of Theorem 6.8 are satisfied. Let μ_q^* be a solution of (6.104) in Theorem 6.8, and set $\gamma_q^* = \frac{\mu_q^*}{q(q-1)}$. We introduce a new function $\Phi_q(\gamma)$ by

$$\Phi_q(\gamma) = \gamma\sigma^2 + \int_{-\infty}^{\infty}\left((1 + (q-1)\gamma(e^x - 1))^{\frac{1}{q-1}}(e^x - 1)\right.$$

$$\left. - x1_{\{|x|\le 1\}}(x)\right)\nu(dx). \qquad (6.122)$$

Then γ_q^* is a solution of the following equation:

$$\Phi_q(\gamma) = \beta \left(= r - b - \frac{1}{2}\sigma^2\right), \qquad (6.123)$$

and (6.102) and (6.103) in Theorem 6.8 are

$$f_q^* = \gamma_q^*\sigma, \qquad (6.124)$$

and

$$e^{g_q^*(x)} = \left(1 + (q-1)\gamma_q^*(e^x - 1)\right)^{\frac{1}{q-1}}. \tag{6.125}$$

(See Remark 6.8.)

From equation (6.123), we formally obtain the following formula:

$$\lim_{q\downarrow 1}\Phi_q(\gamma) = \gamma\sigma^2 + \int_{-\infty}^{\infty}\left(e^{\gamma(e^x-1)}(e^x-1) - x1_{\{|x|\le 1\}(x)}\right)\nu(dx) \equiv \widetilde{\Phi}(\gamma). \tag{6.126}$$

So it is natural to expect that, when $q\downarrow 1$, γ_q^* converges to the solution γ^* of the following equation:

$$\widetilde{\Phi}(\gamma) \equiv \gamma\sigma^2 + \int_{-\infty}^{\infty}(e^{\gamma(e^x-1)}(e^x-1) - x1_{\{|x|\le 1\}}(x))\nu(dx) = \beta. \tag{6.127}$$

We remark here that this equation is just the same as the equation (6.118) in Theorem 6.12 for the MEMM.

Let γ_e^* be the solution of equation (6.127). We can prove that

$$\lim_{q\downarrow 1}\gamma_q^* = \gamma_e^*. \tag{6.128}$$

It is easy to see that $\frac{d\Phi_q}{d\gamma}(\gamma) > 0$ and $\frac{d\widetilde{\Phi}}{d\gamma}(\gamma) > 0$, so the functions $\Phi_q(\gamma)$ and $\widetilde{\Phi}(\gamma)$ are increasing in γ. Therefore the solution γ_q^* of (6.123) and the solution γ_e^* of (6.127) are unique. And from the fact that

$$(1 + (q-1)\gamma(e^x - 1))^{\frac{1}{q-1}} \uparrow e^{\gamma(e^x-1)}, \quad as \quad q\downarrow 1, \tag{6.129}$$

it follows that

$$\Phi_q(\gamma) \to \widetilde{\Phi}(\gamma), \quad as \quad q\downarrow 1. \tag{6.130}$$

From those facts we know that (6.128) holds true. And from (6.128), (6.124), and (6.125), we obtain

$$f_q^* = \gamma_q^*\sigma \quad \to \quad \gamma_e^*\sigma = f_e^* \quad as \quad q\downarrow 1, \tag{6.131}$$

and

$$e^{g_q^*(x)} = \left(1 + (q-1)\gamma_q^*(e^x - 1)\right)^{\frac{1}{q-1}} \quad \to \quad e^{\gamma_e^*(e^x-1)} = e^{g_e^*(x)} \quad as \quad q\downarrow 1. \tag{6.132}$$

So we have obtained the following formula:

$$\lim_{q\downarrow 1}\nu_q^{(ML^qEMM)}(dx) = \lim_{q\downarrow 1}e^{g_q^*(x)}\nu(dx) = e^{\gamma_e^*(e^x-1)}\nu(dx) = \nu^{(MEMM)}. \tag{6.133}$$

Next we analyze the relative entropy $H(P^{(ML^qEMM)}|P^{(MEMM)})$. For the simplicity of notation, we set $P^{(q)} = P^{(ML^qEMM)}$ and $P^* = P^{(MEMM)}$. By the definition of the relative entropy, it holds that

$$H(P^{(q)}|P^*) = E_{P^{(q)}}\left[\log\frac{dP^{(q)}}{dP^*}\right] = E_{P^{(q)}}\left[\log\frac{dP^{(q)}}{dP} - \log\frac{dP^*}{dP}\right]. \quad (6.134)$$

From (6.5) we obtain

$$\log\frac{dP^{(q)}}{dP} = \int_0^T f_q^* dW_s - \frac{1}{2}\int_0^T (f_q^*)^2 ds$$
$$+ \int_0^T \int_{-\infty}^{\infty} g_q^*(x)\,N(ds, dx) - \int_0^T \int_{-\infty}^{\infty} (e^{g_q^*(x)} - 1)\nu(dx)ds,$$

$$(6.135)$$

and

$$\log\frac{dP^*}{dP} = \int_0^T f_e^* dW_s - \frac{1}{2}\int_0^T (f_e^*)^2 ds$$
$$+ \int_0^T \int_{-\infty}^{\infty} g_e^*(x)\,N(ds, dx) - \int_0^T \int_{-\infty}^{\infty} (e^{g_e^*(x)} - 1)\nu(dx)ds.$$

$$(6.136)$$

Therefore

$$\log\frac{dP^{(q)}}{dP^*} = \int_0^T (f_q^* - f_e^*)dW_s - \frac{1}{2}\int_0^T ((f_q^*)^2 - (f_e^*)^2)ds$$
$$+ \int_0^T \int_{-\infty}^{\infty} (g_q^*(x) - g_e^*(x))\,N(ds, dx)$$
$$- \int_0^T \int_{-\infty}^{\infty} (e^{g_q^*(x)} - e^{g_e^*(x)})\nu(dx)ds. \quad (6.137)$$

Set $W_t^{(q)} = \int_0^t \{dW_s - f_q^* ds\}$ and $\tilde{N}^{(q)}(dt, dx) = \{N(dtdx) - e^{g_q^*(x)}\nu(dx)dt\}$, then $W_t^{(q)}$ is a $P^{(q)}$-Wiener process and $\int_0^T \int_{-\infty}^{\infty} \tilde{N}^{(q)}(ds, dx)$ is a $P^{(q)}$-

martingale. Using these facts, we obtain

$$\log \frac{dP^{(q)}}{dP^*} = \int_0^T (f_q^* - f_e^*)(dW_s^{(q)} + f_q^* ds) - \frac{1}{2} \int_0^T ((f_q^*)^2 - (f_e^*)^2) ds$$

$$+ \int_0^T \int_{-\infty}^{\infty} (g_q^*(x) - g_e^*(x)) \, (\widetilde{N}^{(q)}(dt, dx) + e^{g_q^*(x)} \nu(dx) dt)$$

$$- \int_0^T \int_{-\infty}^{\infty} (e^{g_q^*(x)} - e^{g_e^*(x)}) \nu(dx) ds$$

$$= \{P^{(q)} martingale\} + \frac{1}{2} \int_0^T (f_q^* - f_e^*)^2 ds$$

$$+ \int_0^T \int_{-\infty}^{\infty} ((g_q^*(x) - g_e^*(x)) e^{g_q^*(x)} - (e^{g_q^*(x)} - e^{g_e^*(x)})) \nu(dx) ds.$$

$$(6.138)$$

From this it follows that

$$H(P^{(q)}|P^*) = E_{P^{(q)}} \left[\log \frac{dP^{(q)}}{dP^*} \right]$$

$$= T \left\{ \frac{1}{2}(f_q^* - f_e^*)^2 \right.$$

$$\left. + \int_{-\infty}^{\infty} \left((g_q^*(x) - g_e^*(x)) e^{g_q^*(x)} - (e^{g_q^*(x)} - e^{g_e^*(x)}) \right) \nu(dx) \right\}.$$

$$(6.139)$$

Using (6.131) and (6.132), we obtain

$$\lim_{q \downarrow 1} H(P^{(q)}|P^*) = 0. \tag{6.140}$$

Thus we have obtained the following theorem.

Theorem 6.14. *Assume that for any q, $1 < q \leq 2$, the equation*

$$\Phi_q(\gamma) = \beta \ \left(= r - b - \frac{1}{2}\sigma^2 \right) \tag{6.141}$$

has a solution γ_q^*, and that

$$e^{g_q^*(x)} = \left(1 + (q-1)\gamma_q^*(e^x - 1) \right)^{\frac{1}{q-1}} \tag{6.142}$$

is well defined. We also assume that the equation

$$\widetilde{\Phi}(\gamma) = \beta \tag{6.143}$$

has a solution γ_e^*. Then $P^{(ML^q EMM)}$ converges to $P^{(MEMM)}$ as $q \downarrow 1$ in the following senses:

$$\lim_{q \downarrow 1} \nu_q^{(ML^q EMM)}(dx) = \lim_{q \downarrow 1} e^{g_q^*(x)} \nu(dx) = e^{\gamma_e^*(e^x - 1)} \nu(dx) = \nu^{(MEMM)}(dx)$$

$$(6.144)$$

and

$$\lim_{q\downarrow 1} H(P^{(ML^q EMM)}|P^{(MEMM)}) = 0. \qquad (6.145)$$

Remark 6.11. In the above we investigated the case of $q \downarrow 1$. We can discuss the case of $q \uparrow 1$ in a similar way.

Notes

The results of this chapter originated from the arguments of Jeanblanc and Miyahara (2006) [58], and developed from this work. We mention that the statements of this chapter are rather heuristic and not necessarily rigorous. For more rigorous discussions on the subjects of this chapter, see Jeanblanc, Kloeppel, and Miyahara (2007) [57].

There are several papers which discuss subjects similar to that of this chapter, for example [7] and [65].

Chapter 7

The [GLP & MEMM] Pricing Model

We are now ready to explain the [GLP & MEMM] pricing model (geometric Lévy process and minimal entropy martingale measure pricing model). The description of this model is the main subject of this book.

7.1 The Model

We assume that the price process of a bond is deterministic, and that it is given by

$$B_t = \exp\{rt\}, \tag{7.1}$$

where r is a positive constant.

A pricing model consists of the following two parts:
(A) A price process S_t of underlying asset,
(B) A rule to compute the price of option.

If we adopt the martingale measure method for option-pricing, then part (B) of the above is reduced to selecting a suitable martingale measure Q, and then the price of an option X is given by $e^{-rT}E_Q[X]$. The [GLP & MEMM] pricing model is in this framework.

Suppose that a probability space $(\Omega, \mathcal{F}, P)$ and a filtration $\{\mathcal{F}_t, t \geq 0\}$ are given and that the price process $S_t = S_0 e^{Z_t}$ of a stock is defined on this probability space, where Z_t is a Lévy process. We call such a process S_t a geometric Lévy precess (GLP) (see Section 2.2). We assume that $\mathcal{F}_t = \sigma(S_s, 0 \leq s \leq t) = \sigma(Z_s, 0 \leq s \leq t)$ and $\mathcal{F} = \mathcal{F}_T$. Set $\tilde{S}_t = S_t/B_t = S_t e^{-rt}$.

The definition of the minimal entropy martingale measure (MEMM) is given in Section 3 as follows (see Definition 3.3).

Definition 7.1. (MEMM). If an equivalent martingale measure P^* satisfies

$$H(P^*|P) \leq H(Q|P) \qquad \forall Q : \quad equivalent \ martingale \ measure, \quad (7.2)$$

then P^* is called the MEMM of S_t. $H(Q|P)$ is the relative entropy of Q with respect to P, which is defined as

$$H(Q|P) = \left\{ \begin{array}{ll} \int_\Omega \log[\frac{dQ}{dP}]dQ, \ if & Q \ll P, \\ \infty, & otherwise, \end{array} \right\}. \quad (7.3)$$

Remark 7.1. As we have studied in Chapters 4 and 6, the MEMM is obtained as an Esscher-transformed martingale measure by the simple-return process (see Theorem 6.13), and it is also obtained as a minimal distance martingale measure for the distance function $F(x) = x \log x$ (see Section 6.4). We denote the MEMM by $P^{(MEMM)}$ as in previous chapters, and we sometimes denote it by P^* to simplify the notation.

Now we can define the [GLP & MEMM] pricing model.

Definition 7.2. ([GLP & MEMM] Pricing Model). The [GLP & MEMM] pricing model is such a model.
(A) The price process S_t of an underlying asset is a geometric Lévy process (GLP).
(B) The price of an option X is defined to be $e^{-rT}E_{P^*}[X]$, where P^* is the MEMM of S_t.

7.1.1 Sufficient condition for the existence of MEMM

We have already studied the existence problem of the MEMM in Section 6.4 (see Theorems 6.11 and 6.12). Here we state a theorem on the existence of the MEMM, which was obtained in [44], of the original form. Before we state the theorem we need the other expression of S_t

$$S_t = S_0 e^{Z_t} = S_0 \mathcal{E}(\hat{Z})_t, \quad (7.4)$$

where $\mathcal{E}(\hat{Z})_t$ is the Doléans-Dade exponential of $\hat{Z}_t$, and $\hat{Z}_t$ is a Lévy process corresponding to Z_t (see Section 2.3).

Theorem 7.1. (Fujiwara and Miyahara [44] Theorem 3.1).
Suppose that the following condition **(C)** *holds.*
Condition (C)   *There exists $\gamma^* \in R$ which satisfies the following conditions:*

$$(C)_1 \qquad \int_{\{x>1\}} e^x e^{\gamma^*(e^x-1)} \nu(dx) < \infty, \qquad (7.5)$$

$$(C)_2 \qquad b + (\tfrac{1}{2} + \gamma^*)\sigma^2 + \int_{\{|x|>1\}} (e^x - 1)e^{\gamma^*(e^x-1)} \, \nu(dx)$$

$$+ \int_{\{|x|\leq 1\}} \left((e^x - 1)e^{\gamma^*(e^x-1)} - x \right) \nu(dx) = r. \qquad (7.6)$$

And let P^* be the probability measure defined by

$$\left. \frac{dP^*}{dP} \right|_{\mathcal{F}} = \frac{e^{\gamma^* \hat{X}_T}}{E_P[e^{\gamma^* \hat{X}_T}]}. \qquad (7.7)$$

(P^* is the Esscher-transformed probability of P by risk process $\hat{X}_t$.)
    Then the probability measure P^* is well defined and it holds that:
(i) (MEMM): P^* *is the MEMM of S_t.*
(ii) (Lévy process): Z_t *is also a Lévy process w.r.t. P^*, and the generating triplet (A^*, ν^*, b^*) of Z_t under P^* is given by*

$$A^* = \sigma^2, \qquad (7.8)$$

$$\nu^*(dx) = e^{\gamma^*(e^x-1)} \nu(dx), \qquad (7.9)$$

$$b^* = b + \gamma^* \sigma^2 + \int_{R\backslash\{0\}} x 1_{\{|x|\leq 1\}}(x) d(\nu^* - \nu). \qquad (7.10)$$

(iii) *The minimal entropy $H(P^*|P)$ is*

$$H(P^*|P) = T \left\{ \frac{1}{2}\sigma^2 \gamma^{*2} \right.$$

$$\left. + \int_{-\infty}^{\infty} \left(\gamma^* (e^x - 1) e^{\gamma^*(e^x-1)} - e^{\gamma^*(e^x-1)} + 1 \right) \nu(dx) \right\}. \qquad (7.11)$$

Remark 7.2. If the Lévy measure $\nu(dx)$ satisfies the following condition:

$$\int_{\{|x|\leq 1\}} |x|\nu(dx) < \infty, \qquad (7.12)$$

then, by the use of another notation of the generating triplet $(\sigma^2, \nu(dx), b_0)_0$, the condition $(C)_2$ is equivalently expressed as below:

$$b_0 + (\frac{1}{2} + \gamma^*)\sigma^2 + \int_{-\infty}^{\infty} (e^x - 1)e^{\gamma^*(e^x-1)}\nu(dx) = r. \qquad (7.13)$$

In this case, the results of Theorem 7.1 (ii) are equivalently expressed in the following form by the use of the notation $(A^*, \nu^*, b_0^*)_0$:

$$A^* = \sigma^2, \tag{7.14}$$

$$\nu^*(dx) = e^{\gamma^*(e^x - 1)}\nu(dx), \tag{7.15}$$

$$b_0^* = b_0 + \gamma^*\sigma^2. \tag{7.16}$$

These results are the same as the results obtained in Miyahara [86] Theorem 1.

7.1.2   *Properties of geometric Lévy processes under MEMM*

For the calibration of a [GLP & MEMM] pricing model, it is important to specify the class of MEMM-Lévy processes, namely the class of generating triplets (A^*, ν^*, b^*) of Z_t under P^*. (See Chapter 8.)

Set the following:

$$\Phi(\gamma) = b + (\frac{1}{2} + \gamma)\sigma^2 + \int_{\{|x|>1\}} (e^x - 1)e^{\gamma(e^x-1)}\,\nu(dx)$$

$$+ \int_{\{|x|\leq 1\}} \left((e^x - 1)e^{\gamma(e^x-1)} - x\right)\nu(dx). \tag{7.17}$$

Then the condition $(C)_2$ is equivalent to the condition that the following equation

$$\Phi(\gamma) = r \tag{7.18}$$

has a solution γ^*.

It is easy to see that $\Phi(\gamma)$ is a non-decreasing function of γ. (It is an increasing function except for a special case such that $\sigma^2 = 0$ and $\nu(dx) = 0$.) Therefore, if $\Phi(\gamma)$ is continuous and satisfies the following inequality:

$$\lim_{\gamma\to-\infty} \Phi(\gamma) < r < \lim_{\gamma\to\infty} \Phi(\gamma), \tag{7.19}$$

then equation (7.18) has a (unique) solution, and the MEMM exists.

Remark 7.3. In the case where Condition (7.12) is satisfied, the function $\Phi(\gamma)$ is equal to the following function $\Phi_0(\gamma)$:

$$\Phi(\gamma) = \Phi_0(\gamma) = b_0 + (\frac{1}{2} + \gamma)\sigma^2 + \int_{-\infty}^{\infty} (e^x - 1)e^{\gamma(e^x-1)}\,\nu(dx). \tag{7.20}$$

Remark 7.4. (i) If S_t is integrable, then it holds that

$$E[S_t] = S_0 e^{Z_t} = S_0\exp(t\Phi(0)). \tag{7.21}$$

This formula suggests an implication for the value of γ^*. Since S_t is a risky asset, It is natural to assume that

$$\frac{E[S_t]}{S_0} \geq \frac{B_t}{B_0} = e^{rt}. \tag{7.22}$$

Therefore we can usually assume that

$$\Phi(0) \geq r, \tag{7.23}$$

and then the solution γ^* of (7.18) is non-positive.
(ii) Even if S_t is not integrable, it is possible that the solution γ^* of (7.18) is non-positive. Such cases occur very often. In fact, the geometric stable process model is one of such cases. (See Section 7.2.3.) In such a case the Lévy measure $\nu^*(dx)$ of (7.9) is well defined, and the MEMM exists. This is the reason why the MEMM exists for a wide class of geometric Lévy process models.

In the above we have studied the transformation problem of probability measures from the physical probability to the MEMM. The inverse problem shall be discussed in the next chapter. (See Section 8.1.2.)

7.2 Examples of [GLP & MEMM] Pricing Model

In this section we see several examples of [GLP & MEMM] pricing models.

7.2.1 Geometric (Brownian motion + compound Poisson) model (or jump-diffusion model) ([GJD & MEMM])

Suppose that a Lévy process Z_t consists of a continuous part and a compound Poisson part, and that the generating triplet of Z_t is expressed as $(\sigma^2, \lambda\rho(dx), b_0)_0$, where $\sigma^2 > 0$, $\lambda > 0$, and $\rho(dx)$ is a probability measure on $(-\infty, \infty)$ with $\rho(\{0\}) = 0$. Then equation (7.13) for γ^* is

$$b_0 + (\frac{1}{2} + \gamma^*)\sigma^2 + \lambda \int_{(-\infty,\infty)\setminus\{0\}} (e^x - 1)e^{\gamma^*(e^x-1)}\rho(dx) = r. \tag{7.24}$$

Assume that this equation has a solution γ^*, then the process Z_t is also a Lévy process under MEMM P^*, and its generating triplet is $(\sigma^2, \lambda^*\rho^*(dx), b_0 + \gamma^*\sigma^2)_0$, where

$$\lambda^* = \lambda \int_{-\infty}^{\infty} e^{\gamma^*(e^x-1)}\rho(dx), \tag{7.25}$$

and

$$\rho^*(dx) = \frac{e^{\gamma^*(e^x-1)}\rho(dx)}{\int_{-\infty}^{\infty} e^{\gamma^*(e^x-1)}\rho(dx)}. \tag{7.26}$$

1) Brownian Motion Model

Suppose that Z_t consists of a continuous part only. Then Z_t is in the following form:

$$Z_t = bt + \sigma W_t.$$

This model is identical to the Black–Scholes model, and equation (7.24) with $\lambda = 0$ and $b_0 = b$ has a solution $\gamma^* = -\frac{1}{2} - \frac{b-r}{\sigma^2}$, and P^* is the unique martingale measure of S_t.

2) Compound Poisson Model

Suppose that Z_t is a compound Poisson process and that the Lévy measure $\nu(dx)$ of Z_t is given by

$$\nu(dx) = \lambda\rho(dx), \tag{7.27}$$

where λ is a positive constant and $\rho(dx)$ is a probability measure on $(-\infty, \infty)$ such that $\rho(\{0\}) = 0$. Then the equation (7.24) for γ^* is

$$b_0 + \lambda \int_{-\infty}^{\infty} (e^x - 1)e^{\gamma^*(e^x-1)}\rho(dx) = r. \tag{7.28}$$

If this equation has a solution, then the MEMM exists.

As a special case of this example, we can consider the case where the Lévy measure $\nu(dx)$ is discrete, namely in the following form:

$$\nu(dx) = \lambda\rho(dx) = \lambda \sum_{i=1}^{\infty} p_i \delta_{a_i}(dx), \quad p_i > 0, i = 1, 2, \ldots, \quad \sum_{i=1}^{\infty} p_i = 1. \tag{7.29}$$

Then the equation (7.24) for γ^* is

$$b_0 + (\frac{1}{2} + \gamma^*)\sigma^2 + \lambda \sum_{i=1}^{\infty} p_i(e^{a_i} - 1)e^{\gamma^*(e^{a_i}-1)} = r. \tag{7.30}$$

7.2.2 *Geometric variance gamma model* (*[GVG & MEMM]*)

Suppose that Z_t is a variance gamma process. The Lévy measure of Z_t is

$$\nu(dx) = C \left(\frac{1_{\{x<0\}}(x)e^{-c_1|x|} + 1_{\{x>0\}}(x)e^{-c_2|x|}}{|x|} \right) dx, \tag{7.31}$$

where C, c_1, c_2 are constants such that $C > 0$, $c_1, c_2 \geq 0$ and $c_1 + c_2 > 0$. This measure satisfies the condition (7.12). So the generating triplet of Z_t is expressed as $(0, \nu(dx), b_0)_0$, and we can use equation (7.13) for γ. In this case the function $\Phi_0(\gamma)$ in (7.20) is

$$\Phi_0(\gamma) = b_0 + C \left(\int_{-\infty}^{0} \frac{1}{|x|} e^{-c_1|x|} (e^x - 1) e^{\gamma(e^x - 1)} dx \right.$$
$$\left. + \int_{0}^{\infty} \frac{1}{|x|} e^{-c_2|x|} (e^x - 1) e^{\gamma(e^x - 1)} dx \right), \quad (7.32)$$

and the equation for γ^* is

$$\Phi_0(\gamma^*) = r. \quad (7.33)$$

It is easy to see that

$$\Phi_0(\gamma) = \infty \quad if \quad \gamma > 0, \quad (7.34)$$

and that $\Phi_0(\gamma)$ is a continuous increasing function on $(-\infty, 0)$. And it also holds that

$$\lim_{\gamma \to -\infty} \Phi_0(\gamma) = -\infty, \quad (7.35)$$

$$\lim_{\gamma \uparrow 0} \Phi_0(\gamma) = \Phi_0(0) = \begin{cases} = \infty, & if \quad c_2 \leq 1, \\ < \infty, & if \quad c_2 > 1. \end{cases} \quad (7.36)$$

Thus we have obtained the following theorem.

Theorem 7.2. *Let Z_t be a variance gamma process and let the Lévy measure of Z_t be given by (7.31). Then it holds that:*
(1) If $c_2 \leq 1$, then the equation $\Phi_0(\gamma) = r$ has a unique solution γ^*, and the solution is negative. $(\gamma^* < 0)$.
(2) If $c_2 > 1$ and $\Phi_0(0) \geq r$, then the equation $\Phi_0(\gamma) = r$ has a unique solution γ^*, and the solution is non-positive. $(\gamma^* \leq 0)$.
(3) If $c_2 > 1$ and $\Phi_0(0) < r$, then the equation $\Phi_0(\gamma) = r$ has no solution.

When the equation $\Phi_0(\gamma) = r$ has a solution γ^*, then the MEMM of $S_t = S_0 e^{Z_t}$, P^*, exists by Theorem 7.1 and Remark 7.2. And the generating triplet $(A^*, \nu^*, b_0^*)_0$ of Z_t under the MEMM P^* is given by

$$A^* = 0, \quad (7.37)$$

$$\nu^*(dx) = C \left(\frac{1_{\{x<0\}}(x) e^{-c_1|x|} + 1_{\{x>0\}}(x) e^{-c_2|x|}}{|x|} \right) e^{\gamma^*(e^x - 1)} dx, \quad (7.38)$$

$$b_0^* = b_0. \quad (7.39)$$

7.2.3   Geometric stable process model ([GSP & MEMM])

Suppose that Z_t is a stable process and let $(0, \nu(dx), b)$ be its generating triplet. The Lévy measure is given by

$$\nu(dx) = \frac{c_1 1_{\{x<0\}}(x) + c_2 1_{\{x>0\}}(x)}{|x|^{(\alpha+1)}} dx, \qquad (7.40)$$

where $0 < \alpha < 2$, and $c_1 \geq 0$, $c_2 \geq 0$, and $c = c_1 + c_2 > 0$.

Let us examine conditions for the existence of a solution of the equation $\Phi(\gamma) = r$. We consider the case where $c_1 > 0$ and $c_2 > 0$. In this case the function $\Phi(\gamma)$ is

$$\Phi(\gamma) = b + c_1 \int_{-\infty}^{-1} \frac{(e^x - 1)e^{\gamma(e^x-1)}}{|x|^{(\alpha+1)}} dx + c_1 \int_{-1}^{0} \frac{\left((e^x - 1)e^{\gamma(e^x-1)} - x\right)}{|x|^{(\alpha+1)}} dx$$

$$+ c_2 \int_{0}^{1} \frac{\left((e^x - 1)e^{\gamma(e^x-1)} - x\right)}{|x|^{(\alpha+1)}} dx + c_2 \int_{1}^{\infty} \frac{(e^x - 1)e^{\gamma(e^x-1)}}{|x|^{(\alpha+1)}} dx, \qquad (7.41)$$

and

$$\Phi(0) = b + c_1 \int_{-\infty}^{-1} \frac{(e^x - 1)}{|x|^{(\alpha+1)}} dx + c_1 \int_{-1}^{0} \frac{(e^x - 1 - x)}{|x|^{(\alpha+1)}} dx$$

$$+ c_2 \int_{0}^{1} \frac{(e^x - 1 - x)}{|x|^{(\alpha+1)}} dx + c_2 \int_{1}^{\infty} \frac{(e^x - 1)}{|x|^{(\alpha+1)}} dx \qquad (7.42)$$

$$= \infty. \qquad (7.43)$$

It is easily proved that $\Phi(\gamma)$ is a continuous increasing function on $(-\infty, 0)$ and that

$$\lim_{\gamma \to -\infty} \Phi(\gamma) = -\infty, \qquad (7.44)$$

and

$$\lim_{\gamma \uparrow 0} \Phi(\gamma) = \infty. \qquad (7.45)$$

Therefore the equation $\Phi(\gamma) = r$ has a unique negative solution γ^*. Thus we have obtained the following result.

Theorem 7.3. Let Z_t be a stable process and let the Lévy measure of Z_t be given by (7.40). Then it holds that, under the assumption $c_1, c_2 > 0$, the equation $\Phi(\gamma) = r$ has a unique solution γ^*, and the solution γ^* is negative. ($\gamma^* < 0$.)

Remark 7.5. In a case where at least one of c_1 and c_2 is equal to 0, the existence of the solution γ^* is dependent on α and b. (See [44] Example 3.2.)

If the equation $\Phi(\gamma) = r$ has a solution γ^*, then, by Theorem 7.1 (or Theorem 6.11 and Theorem 6.12), the MEMM P^* of $S_t = S_0 e^{Z_t}$ exists and the generating triplet $(A^*, \nu^*(dx), b^*)$ of Z_t under P^* is

$$A^* = A = 0, \tag{7.46}$$

$$\nu^*(dx) = e^{\gamma^*(e^x-1)}\nu(dx)$$

$$= \left(\frac{c_1 1_{\{x<0\}}(x) + c_2 1_{\{x>0\}}(x)}{|x|^{(\alpha+1)}} \right) e^{\gamma^*(e^x-1)}dx, \tag{7.47}$$

$$b^* = b + \int_{-\infty}^{\infty} x 1_{\{|x|\leq 1\}}(x)d(\nu^* - \nu). \tag{7.48}$$

As a corollary of the above theorem, we get the finiteness of moments of S_t under the MEMM. Actually the moments $E_{P^*}[|S_t|^k]$, $k = 1, 2, \ldots$, are calculated as below:

$$E_{P^*}[|S_t|^k] = E_{P^*}[e^{kZ_t}] = \phi^*_{Z_t}(-ik)$$

$$= \exp\left\{ t \left(\int_{-\infty}^{\infty} (e^{kx} - 1 - k 1_{\{|x|\leq 1\}}(x))\nu^*(dx) + bk \right) \right\}$$

$$= \exp\left\{ t \left(\int_{-\infty}^{\infty} (e^{kx} - 1 - k 1_{\{|x|\leq 1\}}(x))e^{\gamma^*(e^x-1)}\nu(dx) + bk \right) \right\}.$$

$$\tag{7.49}$$

From the fact that γ^* is negative, it is easy to see that the integral of the above formula converges, and we can obtain the following corollary:

Corollary 7.1. *When $c_1, c_2 > 0$, any moment of $S_t = S_0 e^{Z_t}$ under the MEMM P^*, $E_{P^*}[|S_t|^k]$, $k = 1, 2, \ldots$ is finite.*

Remark 7.6. Under the original measure P, $S_t, t > 0$, is not integrable. But under the MEMM P^*, any moment $E_{P^*}[|S_t|^k], k = 1, 2, \ldots$, of S_t is finite. This property of a geometric stable process model is very useful when we apply this model to option-pricing problems. In fact, if the option O satisfies such a condition as $|O| < a(S_T)^k$ for some constants $a > 0$ and $k > 0$, then the price of O is computable as the expectation $E_{P^*}[Oe^{-rT}]$.

7.2.4 Geometric CGMY model ([GCGMY & MEMM])

Suppose that Z_t is a CGMY process. Then the Lévy measure of Z_t is

$$\nu(dx) = C \left(\frac{1_{\{x<0\}}(x)e^{-G|x|} + 1_{\{x>0\}}(x)e^{-M|x|}}{|x|^{(1+Y)}} \right) dx, \qquad (7.50)$$

where $C > 0, G \geq 0, M \geq 0, Y < 2$. If $Y \leq 0$, then it is assumed that $G > 0$ and $M > 0$. We mention here that the case of $Y = 0$ is the variance gamma process case, and the case such that $G = M = 0$ and $0 < Y < 2$ is the symmetric stable process case. In the sequel we assume that $G, M > 0$.
In this case the function $\Phi(\gamma)$ defined in (7.17) is

$$\begin{aligned}
\Phi(\gamma) = b + C \Bigg(&\int_{-\infty}^{-1} \frac{(e^x - 1)e^{\gamma(e^x-1)}e^{-G|x|}}{|x|^{(1+Y)}} dx \\
&+ \int_{-1}^{0} \frac{\left((e^x - 1)e^{\gamma(e^x-1)} - x)\right) e^{-G|x|}}{|x|^{(1+Y)}} dx \\
&+ \int_{0}^{1} \frac{\left((e^x - 1)e^{\gamma(e^x-1)} - x)\right) e^{-M|x|}}{|x|^{(1+Y)}} dx \\
&+ \int_{1}^{\infty} \frac{(e^x - 1)e^{\gamma(e^x-1)}e^{-M|x|}}{|x|^{(1+Y)}} dx \Bigg),
\end{aligned} \qquad (7.51)$$

and $\Phi(0)$ is

$$\begin{aligned}
\Phi(0) = b + C \Bigg(&\int_{-\infty}^{-1} \frac{(e^x - 1)e^{-G|x|}}{|x|^{(1+Y)}} dx \\
&+ \int_{-1}^{0} \frac{(e^x - 1 - x)e^{-G|x|}}{|x|^{(1+Y)}} dx + \int_{0}^{1} \frac{(e^x - 1 - x)e^{-M|x|}}{|x|^{(1+Y)}} dx \\
&+ \int_{1}^{\infty} \frac{(e^x - 1)e^{-M|x|}}{|x|^{(1+Y)}} dx \Bigg).
\end{aligned} \qquad (7.52)$$

We can carry on a similar calculation to the one above for the variance gamma process and the stable process, and obtain the following theorem.

Theorem 7.4. *Let Z_t be a CGMY process and let the Lévy measure of Z_t be given by (7.50). Then it holds that:*
(1) If $0 < M \leq 1$, then the equation $\Phi(\gamma) = r$ has a unique solution γ^*, and the solution is negative. ($\gamma^* < 0$.)
(2) If $M > 1$ and $\Phi(0) \geq r$, then the equation $\Phi(\gamma) = r$ has a unique solution γ^*, and the solution is non-positive. ($\gamma^* \leq 0$.)
(3) If $M > 1$ and $\Phi(0) < r$, then the equation $\Phi(\gamma) = r$ has no solution.

When the equation $\Phi(\gamma) = r$ has a solution γ^*, then the MEMM P^* of $S_t = S_0 e^{Z_t}$ exists and the generating triplet $(A^*, \nu^*(dx), b^*)$ of Z_t under P^* is

$$A^* = A = 0, \tag{7.53}$$

$$\nu^*(dx) = C \left(\frac{1_{\{x<0\}}(x)e^{-G|x|} + 1_{\{x>0\}}(x)e^{-M|x|}}{|x|^{(1+Y)}} \right) e^{\gamma^*(e^x - 1)} dx, \tag{7.54}$$

$$b^* = b + \int_{-\infty}^{\infty} x 1_{\{|x| \leq 1\}}(x) d(\nu^* - \nu). \tag{7.55}$$

7.3 Why the Geometric Lévy Process?

There are many studies on the distribution of log return of a stock (see [36], [54], and [80] for example). Many researchers have insisted that the distribution of log return of a stock is not normal, saying that the distribution of log return is asymmetric and has a fat tail property. The geometric Lévy process fits to these results.

7.4 Why the MEMM?

A geometric Lévy process model is an incomplete market model, and so there are many equivalent martingale measures for this model. Therefore we have to select a suitable martingale measure among them for option pricing.

We have already seen some examples of the equivalent martingale measures: Esscher-transformed martingale measures (ESSTMM) in Chapter 4 and minimal distance martingale measures (MDMM) in Chapter 6. Among them the minimal entropy martingale measure (MEMM) is a special one in the sense that the MEMM is obtained as an Esscher-transformed martingale measure by the simple-return process and it is also obtained as a MDMM corresponding to the entropy distance. (See Chapters 4 and 6.)

We will see below that the MEMM has many good properties, and we insist that this martingale measure is the most desirable candidate among all known martingale measures for a suitable equivalent martingale measure.

1) Relationship to Kullback–Leibler Information Number

The relative entropy is very popular in the field of information theory, and it is called Kullback–Leibler information number (see [56] p.23) or

Kullback–Leibler distance (see [27] p.18). In the sense of Kullback–Leibler distance, the MEMM is the nearest equivalent martingale measure to the original probability P.

2) Relationship to Large Deviation Theory

The relative entropy is very strongly related to large deviation theory. Sanov's theorem (see [27] pp.291–304 or [56] pp.110–111) tells us that the MEMM is the most possible empirical probability measure of paths of price process in the class of all equivalent martingale measures. In this sense the MEMM is an exceptional martingale measure.

3) Relationship to Esscher Transformation

The Esscher transform is very popular in the field of risk management. We have seen in Section 6.4 that the MEMM is identified with the simple-return Esscher-transformed martingale measure.

4) Applicability to Geometric Stable Model

The geometric stable process model is considered to be one of the most important models for a price process, and has been studied by many researchers. (See [36], [77], [33], and [103].) As we see in the next section, the MEMM method can be applied to geometric stable process models. We emphasize here that the MEMM is the only one martingale measure which can be applied to the geometric stable process model among the frequently used martingale measures.

5) Economical Characterization of MEMM by Exponential Utility Function

We consider the following exponential utility function:

$$u_\alpha(x) = 1 - e^{-\alpha x}, \tag{7.56}$$

and set

$$J_\alpha(c, B) := \sup_{\theta \in \Theta} E_P[u_\alpha(c + G(\theta)_T - B)]$$
$$= \sup_{\theta \in \Theta} E_P[1 - \exp\{-\alpha(c + G(\theta) - B)\}], \tag{7.57}$$

where Γ is a set of strategies, $G(\theta)$ is the gain of a strategy θ, and B is a contingent claim.

This quantity $J_\alpha(c, B)$ is related to the relative entropy by the following duality relation (see [29] and [6]):

$$J_\alpha(c, B) = 1 - \exp\left\{-\inf_{Q \in \mathcal{M}}(H(Q|P) + \alpha c - E_Q[\alpha B])\right\}$$

$$= 1 - e^{-\alpha c}\exp\left\{\alpha \sup_{Q \in \mathcal{M}}\left(E_Q[B] - \frac{1}{\alpha}H(Q|P)\right)\right\}, \quad (7.58)$$

where $\mathcal{M}$ is a convex set of local martingale measures corresponding to Γ.

Using this duality property, we can characterize the MEMM price $E_{P^*}[B]$. We give the definition of utility indifference price $p_\alpha(c, B)$. (See [29] Section 4.2, [51], and [52].)

Definition 7.3. The value $p_\alpha(c, B)$ which satisfies the equation

$$J_\alpha(c + p_\alpha(c, B), B) = J_\alpha(c, 0) \quad (7.59)$$

is called the utility indifference price of B.

It is easy to see that the value $p_\alpha(c, B)$ does not depend on c, so we use a simplified notation $p_\alpha(B)$. The parameter α is called a risk-sensitivity parameter.

In Section 4 of [44] the following result is obtained.

Theorem 7.5. (Fujiwara and Miyahara [44] Corollary 4.1).
Suppose that the duality relation (7.58) holds. Let $p_\alpha(B)$ be the utility indifference price of a bounded contingent claim B for the exponential utility function $u_\alpha(x)$. Then it holds that:
(i) $p_\alpha(B) \geq E_{P^*}[B]$ *for any $\alpha > 0$,*
(ii) *If $0 < \alpha < \beta$, then $p_\alpha(B) < p_\beta(B)$,*
(iii) $\lim_{\alpha \downarrow 0} p_\alpha(B) = E_{P^*}[B]$.

From this theorem we can say that the MEMM price $E_{P^*}[B]$ is the lower bound of the price of B at which price a risk-aversive agent buys B.

6) Volatility Smile/Smirk Property
The volatility smile/smirk property is supposed to be a very important property which a good pricing model should have. Computer simulation results show us that the [GLP & MEMM] pricing models have this property. We shall see this subject in Section 8.2. (See also [93].)

From the facts 1–6 described above, the MEMM is shown as an exceptional measure in the class of all equivalent martingale measures.

7.5   Comparison of Equivalent Martingale Measures for Geometric Lévy Processes

As we have seen in the previous chapters, there are many equivalent martingale measures for a geometric Lévy process. In this section we see the differences between those equivalent martingale measures.

First it should be noticed that a Lévy process Z_t is also a Lévy process under the equivalent martingale measures obtained in previous chapters.

7.5.1   *Corresponding risk process relating to Esscher-transformed MM*

As we have described in Chapter 4, each Esscher-transformed martingale measure corresponds to a risk process which determines the Esscher transformation. We summarize the known results of this correspondence below. (See Chapter 4.)

• MEMM $P^{(MEMM)}$: To the minimal entropy martingale measure, the corresponding risk process is the simple-return process $\tilde{Z}_t$.

• ESSMM $P^{(ESSMM)}$: To the Esscher martingale measure, the corresponding risk process is the compound-return process Z_t.

• MCMM $P^{(MCMM)}$: To the mean correcting martingale measure, the corresponding risk process is the Wiener process (Gawssian part) W_t.

7.5.2   *Integrability condition of Lévy measures for the existence of martingale measures*

Conditions for the existence of each equivalent martingale measure have been obtained in the previous chapters. We summarize the known results below. (See Chapters 4 and 6.)

• MEMM $P^{(MEMM)}$: The condition

$$\int_{\{|x|>1\}} |(e^x - 1)| e^{\gamma^*(e^x - 1)} \, \nu(dx) < \infty \qquad (7.60)$$

is necessary for the existence of the MEMM, where γ^* is a solution of the following equation:

$$b + (\tfrac{1}{2} + \gamma^*)\sigma^2 + \int_{\{|x|>1\}} (e^x - 1)e^{\gamma^*(e^x - 1)} \, \nu(dx)$$
$$+ \int_{\{|x|\leq 1\}} \left((e^x - 1)e^{\gamma^*(e^x - 1)} - x \right) \nu(dx) = r. \qquad (7.61)$$

Remark 7.7. If the solution of (7.61) is negative ($\gamma^* < 0$), then the above condition (7.60) is satisfied for a wide class of Lévy measures. And the case where the solution γ^* of the equation (7.61) is negative occurs very often.

- ESSMM $P^{(ESSMM)}$: The condition

$$\int_{\{|x|>1\}} |e^x - 1|\, e^{h^* x} \nu(dx) < \infty \qquad (7.62)$$

is necessary for the existence of the ESSMM, where h^* is a solution of

$$b + (\tfrac{1}{2} + h)\sigma^2 + \int_{\{|x|\leq 1\}} \left((e^x - 1)e^{hx} - x \right) \nu(dx)$$
$$+ \int_{\{|x|>1\}} (e^x - 1)e^{hx}\, \nu(dx) = r. \qquad (7.63)$$

- MCMM $P^{(MCMM)}$: The condition

$$\sigma > 0, \quad \text{and} \quad \int_{\{|x|>1\}} |\,(e^x - 1)\,|\nu(dx) < \infty \qquad (7.64)$$

is necessary for the existence of the MCMM.

- MVEMM $P^{(MVEMM)}$: The condition

$$\int_{\{|x|>1\}} (e^x - 1)^2 \nu(dx) < \infty \qquad (7.65)$$

is necessary for the existence of the MVEMM.

- ML^{q}EMM $P^{(ML^q EMM)}$, $q \neq 0, 1$: The condition

$$\int_{\{|x|>1\}} \left(1 + \frac{\gamma_q^*(e^x - 1)}{q} \right)^{\frac{1}{q-1}} (e^x - 1)\nu(dx) < \infty \qquad (7.66)$$

is necessary for the existence of the ML^{q}EMM, where γ_q^* is a solution of

$$\frac{\gamma_q \sigma^2}{q(q - 1)} + \int_{-\infty}^{\infty} \left(\left(1 + \frac{\gamma_q(e^x - 1)}{q} \right)^{\frac{1}{q-1}} (e^x - 1) - x 1_{\{|x|\leq 1\}}(x) \right) \nu(dx) = \beta. \qquad (7.67)$$

Remark 7.8. (See Section 6.5.) Set $\gamma_q^* = \frac{\gamma_q^*}{q(q-1)}$ and define

$$\Phi_q(\gamma) = \gamma \sigma^2 + \int_{-\infty}^{\infty} \left((1 + (q - 1)\gamma(e^x - 1))^{\frac{1}{q-1}} (e^x - 1) \right.$$
$$\left. - x 1_{\{|x|\leq 1\}}(x) \right) \nu(dx). \qquad (7.68)$$

Then γ_q^* is a solution of the following equation:

$$\Phi_q(\gamma) = \beta \ (= r - b - \tfrac{1}{2}\sigma^2), \qquad (7.69)$$

and the formulas (6.102) and (6.103) in Theorem 6.8 are in the following forms:

$$f_q^* = \gamma_q^* \sigma \tag{7.70}$$

and

$$e^{g_q^*(x)} = \left(1 + (q-1)\gamma_q^*(e^x - 1)\right)^{\frac{1}{q-1}}. \tag{7.71}$$

Using these notations, the condition (7.66) is expressed in the following form:

$$\int_{\{|x|>1\}} (e^x - 1)e^{g_q^*(x)}\nu(dx) < \infty. \tag{7.72}$$

This is a similar form to (7.60) and (7.62).

7.5.3 *Corresponding utility function*

From the discussions in Chapter 5, we know the following relations.

- MEMM $P^{(MEMM)}$: This measure corresponds to the exponential utility function:

$$u(x) = 1 - e^{-\alpha x}. \tag{7.73}$$

- MVEMM $P^{(MVEMM)}$: This measure corresponds to the following utility function:

$$u(x) = -x^2. \tag{7.74}$$

- ML^qEMM $P^{(ML^q EMM)}$, $q \neq 0, 1$: This measure corresponds to the power utility function

$$u(x) = \frac{1}{p}|x|^p, \ p = \frac{q}{q-1}, \ \text{if} \ p < 1, \ p \neq 0 \ (\Leftrightarrow q < 1, \ q \neq 0), \tag{7.75}$$

or

$$u(x) = \frac{-1}{p}|x|^p, \ p = \frac{q}{q-1}, \ \text{if} \ p > 1 \ (\Leftrightarrow q > 1). \tag{7.76}$$

7.6   The Explicit Form of Lévy Measure of Z_t under an Equivalent Martingale Measure

In this section we give explicit forms of equivalent martingale measures for several geometric Lévy processes.

Let $(\sigma^2, \nu(dx), b)$ be the generating triplet of a Lévy process Z_t under the original measure P, and let $(A^*, \nu^*(dx), b^*)$ be the generating triplet of Z_t under an equivalent martingale measure Q. Then it always holds that $A^* = \sigma^2$. And if we know a new Lévy measure $\nu^*(dx)$, the parameter b^* is determined easily from the martingale condition for $S_t = S_0 e^{Z_t}$. Therefore the main problem for the determination of $(A^*, \nu^*(dx), b^*)$ is reduced to the determination of the Lévy measure $\nu^*(dx)$.

7.6.1   *General form of Lévy measure of Z_t under equivalent martingale measures*

We summarize the general form of Lévy measures of Z_t under equivalent martingale measures as follows.

• MEMM: If the MEMM exists, then the Lévy measure of Z_t under the MEMM is given by

$$\nu^{(MEMM)}(dx) = e^{\gamma^*(e^x - 1)}\nu(dx), \qquad (7.77)$$

where γ^* is a solution of

$$b + (\frac{1}{2} + \gamma)\sigma^2 + \int_{-\infty}^{\infty} \left((e^x - 1)e^{\gamma(e^x - 1)} - x1_{\{|x| \leq 1\}}(x) \right) \nu(dx) = r. \quad (7.78)$$

We remark here that the solution of this equation γ^* is usually negative.

• ESSMM: If the ESSMM exists, then the Lévy measure of Z_t under the ESSMM is given by

$$\nu^{(ESSMM)}(dx) = e^{h^*x}\nu(dx), \qquad (7.79)$$

where h^* is a solution of

$$b + (\frac{1}{2} + h)\sigma^2 + \int_{-\infty}^{\infty} \left((e^x - 1)e^{hx} - x1_{\{|x| \leq 1\}}(x) \right) \nu(dx) = r. \quad (7.80)$$

• MCMM: If $\sigma^2 \neq 0$, then the MCMM exists and the Lévy measure of Z_t under the MCMM is unchanged:

$$\nu^{(MCMM)}(dx) = \nu(dx). \qquad (7.81)$$

- MVEMM: If the MVEMM exists, then the Lévy measure of Z_t under the MVEMM is given by

$$\nu^{(MVEMM)}(dx) = (1 + \gamma^*(e^x - 1))\,\nu(dx), \qquad (7.82)$$

where γ^* is the solution of the following equation:

$$b + \left(\frac{1}{2} + \gamma\right)\sigma^2 + \int_{-\infty}^{\infty} \left((1 + \gamma(e^x - 1))(e^x - 1) - x1_{\{|x| \le 1\}}(x)\right)\nu(dx) = r. \qquad (7.83)$$

- ML^qEMM: If the ML^qEMM exists, then the Lévy measure of Z_t under the ML^qEMM is given by

$$\nu^{(ML^qEMM)}(dx) = e^{g_q^*(x)}\nu(dx) = \left(1 + (q-1)\gamma_q^*(e^x - 1)\right)^{\frac{1}{q-1}}\nu(dx), \qquad (7.84)$$

where γ_q^* is the solution of

$$b + \left(\frac{1}{2} + \gamma\right)\sigma^2 + \int_{-\infty}^{\infty} \left((1 + (q-1)\gamma(e^x - 1))^{\frac{1}{q-1}}(e^x - 1)\right.$$
$$\left. - x1_{\{|x| \le 1\}}(x)\right)\nu(dx) = r. \qquad (7.85)$$

(See Remark 7.8.)

7.6.2 *Geometric variance gamma model*

The Lévy measure of a variance gamma process is

$$\nu(dx) = C\frac{\left(1_{\{x<0\}}(x)e^{-c_1|x|} + 1_{\{x>0\}}(x)e^{-c_2|x|}\right)}{|x|}dx, \qquad (7.86)$$

where $C > 0$, $c_1 \ge 0$, $c_2 \ge 0$, and $c_1 + c_2 > 0$.

- MEMM: If the equation

$$b + \int_{-\infty}^{\infty} \left((e^x - 1)e^{\gamma(e^x - 1)} - x1_{\{|x| \le 1\}}(x)\right)\nu(dx) = r, \qquad (7.87)$$

has a solution γ^*, then the MEMM exists, and the Lévy measure of Z_t under the MEMM is

$$\nu^{(MEMM)}(dx) = e^{\gamma^*(e^x - 1)}\nu(dx)$$
$$= C\frac{\left(1_{\{x<0\}}(x)e^{-c_1|x|} + 1_{\{x>0\}}(x)e^{-c_2|x|}\right)}{|x|}e^{\gamma^*(e^x - 1)}dx. \qquad (7.88)$$

Conditions for the existence are as follows (see Theorem 7.2):

(1) $c_2 \leq 1$,

or

(2) $c_2 > 1 +$ (some additional conditions).

- ESSMM: If the equation

$$b + \int_{-\infty}^{\infty} \left((e^x - 1)e^{hx} - x1_{\{|x| \leq 1\}}(x) \right) \nu(dx) = r, \qquad (7.89)$$

has a solution h^*, then the ESSMM exists, and the Lévy measure of Z_t under the ESSMM is given by

$$\nu^{(ESSMM)}(dx) = e^{h^* x} \nu(dx)$$

$$= C \frac{\left(1_{\{x<0\}}(x)e^{-c_1|x|} + 1_{\{x>0\}}(x)e^{-c_2|x|} \right)}{|x|} e^{h^* x} dx. \qquad (7.90)$$

In this case, h^* must be in $(-c_1, c_2)$.

- MCMM: Since $\sigma^2 = 0$, the MCMM does not exist.

- MVEMM: If the equation

$$b + \int_{-\infty}^{\infty} \left((1 + \gamma(e^x - 1))(e^x - 1) - x1_{\{|x| \leq 1\}}(x) \right) \nu(dx) = r, \qquad (7.91)$$

has a solution γ^*, then the MVEMM exists, and the Lévy measure of Z_t under the MVEMM is

$$\nu^{(MVEMM)}(dx) = (1 + \gamma^*(e^x - 1)) \nu(dx)$$

$$= C \frac{\left(1_{\{x<0\}}(x)e^{-c_1|x|} + 1_{\{x>0\}}(x)e^{-c_2|x|} \right)}{|x|} (1 + \gamma^*(e^x - 1)) dx. \quad (7.92)$$

In this case, γ^* must be in $(0,1)$.

- ML^qEMM: If the equation

$$b + \int_{-\infty}^{\infty} \left((1 + (q-1)\gamma(e^x - 1))^{\frac{1}{q-1}}(e^x - 1) - x1_{\{|x| \leq 1\}}(x) \right) \nu(dx) = r, \qquad (7.93)$$

has a solution γ_q^*, then the ML^qEMM exists, and the Lévy measure of Z_t under the ML^qEMM is

$$\nu^{(ML^q EMM)}(dx) = \left(1 + (q-1)\gamma_q^*(e^x - 1) \right)^{\frac{1}{q-1}} \nu(dx)$$

$$= C \frac{\left(1_{\{x<0\}}(x)e^{-c_1|x|} + 1_{\{x>0\}}(x)e^{-c_2|x|} \right)}{|x|} \left(1 + (q-1)\gamma_q^*(e^x - 1) \right)^{\frac{1}{q-1}} dx.$$

$$(7.94)$$

7.6.3 Geometric stable process model

The Lévy measure of a stable process is

$$\nu(dx) = \frac{\left(c_1 1_{\{x<0\}}(x) + c_2 1_{\{x>0\}}(x)\right)}{|x|^{\alpha+1}} dx, \qquad (7.95)$$

where $0 < \alpha < 2$, $c_1 \geq 0$, $c_2 \geq 0$, and $c_1 + c_2 > 0$.

• MEMM: If the equation

$$b + \int_{-\infty}^{\infty} \left((e^x - 1)e^{\gamma(e^x - 1)} - x 1_{\{|x| \leq 1\}}(x) \right) \nu(dx) = r, \qquad (7.96)$$

has a solution γ^*, then the MEMM exists, and the Lévy measure of Z_t under the MEMM is

$$\nu^{(MEMM)}(dx) = e^{\gamma^*(e^x - 1)}\nu(dx)$$

$$= \frac{\left(c_1 1_{\{x<0\}}(x) + c_2 1_{\{x>0\}}(x)\right)}{|x|^{\alpha+1}} e^{\gamma^*(e^x - 1)} dx. \qquad (7.97)$$

A sufficient condition for the existence of the MEMM is that $c_1 > 0$ and $c_2 > 0$. (See Theorem 7.3.)

• Other martingale measures (ESSMM, MCMM, MVEMM, ML^qEMM) do not exist in general.

7.6.4 Geometric CGMY model

The Lévy measure of a CGMY process is

$$\nu(dx) = C\frac{\left(1_{\{x<0\}}(x)\exp(-G|x|) + 1_{\{x>0\}}(x)\exp(-M|x|)\right)}{|x|^{1+Y}} dx, \qquad (7.98)$$

where $C > 0$, $G \geq 0$, $M \geq 0$, and $Y < 2$.

• MEMM: If the equation

$$b + \int_{-\infty}^{\infty} \left((e^x - 1)e^{\gamma(e^x - 1)} - x 1_{\{|x| \leq 1\}}(x) \right) \nu(dx) = r, \qquad (7.99)$$

has a solution γ^*, then the MEMM exists, and the Lévy measure of Z_t under the MEMM is

$$\nu^{(MEMM)}(dx) = e^{\gamma^*(e^x - 1)}\nu(dx)$$

$$= C\frac{\left(1_{\{x<0\}}(x)\exp(-G|x|) + 1_{\{x>0\}}(x)\exp(-M|x|)\right)}{|x|^{1+Y}} e^{\gamma^*(e^x - 1)} dx.$$

$$\qquad (7.100)$$

Conditions for the existence of MEMM are given in Theorem 7.4.

- ESSMM: If the equation

$$b + \int_{-\infty}^{\infty} \left((e^x - 1)e^{hx} - x1_{\{|x|\leq 1\}}(x) \right) \nu(dx) = r, \qquad (7.101)$$

has a solution h^*, then the ESSMM exists, and the Lévy measure of Z_t under the ESSMM is given by

$$\nu^{(ESSMM)}(dx) = e^{h^* x}\nu(dx)$$
$$= C\frac{\left(1_{\{x<0\}}(x)\exp(-G|x|) + 1_{\{x>0\}}(x)\exp(-M|x|)\right)}{|x|^{1+Y}}e^{h^* x}dx.$$
$$(7.102)$$

In this case, h^* must be in $(-G, M)$.

- MCMM: Since $\sigma^2 = 0$, the MCMM does not exist.

- MVEMM: If the equation

$$b + \int_{-\infty}^{\infty} \left((1 + \gamma(e^x - 1))(e^x - 1) - x1_{\{|x|\leq 1\}}(x) \right) \nu(dx) = r, \qquad (7.103)$$

has a solution γ^*, then the MVEMM exists, and the Lévy measure of Z_t under the MVEMM is

$$\nu^{(MVEMM)}(dx) = (1 + \gamma^*(e^x - 1))\nu(dx)$$
$$= C\frac{\left(1_{\{x<0\}}(x)\exp(-G|x|) + 1_{\{x>0\}}(x)\exp(-M|x|)\right)}{|x|^{1+Y}}(1 + \gamma^*(e^x - 1))dx.$$
$$(7.104)$$

In this case, γ^* must be in (0,1).

- ML^qEMM: If the equation

$$b + \int_{-\infty}^{\infty} \left((1 + (q-1)\gamma(e^x - 1))^{\frac{1}{q-1}}(e^x - 1) - x1_{\{|x|\leq 1\}}(x) \right) \nu(dx) = r, \qquad (7.105)$$

has a solution γ_q^*, then the ML^qEMM exists, and the Lévy measure of Z_t under the ML^qEMM is

$$\nu^{(ML^q EMM)}(dx) = \left(1 + (q-1)\gamma_q^*(e^x - 1)\right)^{\frac{1}{q-1}}\nu(dx)$$
$$= C\frac{\left(1_{\{x<0\}}(x)\exp(-G|x|) + 1_{\{x>0\}}(x)\exp(-M|x|)\right)}{|x|^{1+Y}}$$
$$\times \left(1 + (q-1)\gamma_q^*(e^x - 1)\right)^{\frac{1}{q-1}}dx. \qquad (7.106)$$

7.6.5   *Merton model (jump-diffusion model)*

The generating triplet of Z_t of the Merton model is expressed as $(\sigma^2, \lambda\rho(dx), b_0)_0$, where $\sigma^2 > 0$, $\lambda > 0$, and $\rho(dx)$ is a probability measure on $(-\infty, \infty)$ with $\rho(0) = 0$.

• MEMM: If the equation

$$b_0 + (\frac{1}{2} + \gamma^*)\sigma^2 + \lambda \int_{-\infty}^{\infty} (e^x - 1)e^{\gamma(e^x-1)}\rho(dx) = r, \qquad (7.107)$$

has a solution γ^*, then the MEMM exists, and the Lévy measure of Z_t under the MEMM is

$$\nu^{(MEMM)}(dx) = e^{\gamma^*(e^x-1)}\nu(dx) = \lambda^*\rho^*(dx), \qquad (7.108)$$

where

$$\lambda^* = \lambda \int_{-\infty}^{\infty} e^{\gamma^*(e^x-1)}\rho(dx), \qquad (7.109)$$

and

$$\rho^*(dx) = \frac{e^{\gamma^*(e^x-1)}\rho(dx)}{\int_{-\infty}^{\infty} e^{\gamma^*(e^x-1)}\rho(dx)}. \qquad (7.110)$$

And the generating triplet of Z_t under the MEMM is

$$(\sigma^2, \lambda^*\rho^*(dx), b_0 + \gamma^*\sigma^2)_0. \qquad (7.111)$$

• ESSMM: If the equation

$$b_0 + \left(\frac{1}{2} + h\right)\sigma^2 + \lambda \int_{-\infty}^{\infty} \left((e^x - 1)e^{hx}\right)\rho(dx) = r, \qquad (7.112)$$

has a solution h^*, then the ESSMM exists, and the Lévy measure of Z_t under the ESSMM is given by

$$\nu^{(ESSMM)}(dx) = e^{h^*x}\nu(dx) = e^{h^*x}\lambda\rho(dx) = \lambda^{**}\rho^{**}(dx), \quad (7.113)$$

where

$$\lambda^{**} = \lambda \int_{-\infty}^{\infty} e^{h^*x}\rho(dx), \qquad (7.114)$$

and

$$\rho^{**}(dx) = \frac{e^{h^*x}\rho(dx)}{\int_{-\infty}^{\infty} e^{h^*x}\rho(dx)}. \qquad (7.115)$$

• MCMM: Since $\sigma^2 > 0$, the MCMM exists.

• MVEMM: If the equation

$$b_0 + \left(\frac{1}{2} + \gamma\right)\sigma^2 + \lambda \int_{-\infty}^{\infty} (1 + \gamma(e^x - 1))(e^x - 1)\rho(dx) = r, \quad (7.116)$$

has a solution γ^*, then the MVEMM exists, and the Lévy measure of Z_t under the MVEMM is given by

$$\nu^{(MVEMM)}(dx) = (1 + \gamma^*(e^x - 1))\lambda\rho(dx) = \lambda^{***}\rho^{***}(dx), \quad (7.117)$$

where

$$\lambda^{***} = \lambda \int_{-\infty}^{\infty} (1 + \gamma^*(e^x - 1))\rho(dx), \quad (7.118)$$

and

$$\rho^{***}(dx) = \frac{(1 + \gamma^*(e^x - 1))\rho(dx)}{\int_{-\infty}^{\infty}(1 + \gamma^*(e^x - 1))\rho(dx)}. \quad (7.119)$$

• ML^qEMM: If the equation

$$b_0 + \left(\frac{1}{2} + \gamma_q\right)\sigma^2 + \lambda \int_{-\infty}^{\infty} (1 + (q-1)\gamma_q)^{\frac{1}{q-1}}(e^x - 1)\rho(dx) = r, \quad (7.120)$$

has a solution γ_q^*, then the ML^qEMM exists, and the Lévy measure of Z_t under the ML^qEMM is given by

$$\nu^{(ML^qEMM)}(dx) = (1 + (q-1)\gamma_q)^{\frac{1}{q-1}}\lambda\rho(dx) = \lambda_q^{****}\rho_q^{****}(dx), \quad (7.121)$$

where

$$\lambda_q^{****} = \lambda \int_{-\infty}^{\infty} \left(1 + (q-1)\gamma_q^*(e^x - 1)\right)^{\frac{1}{q-1}}\rho(dx), \quad (7.122)$$

and

$$\rho_q^{****}(dx) = \frac{\left(1 + (q-1)\gamma_q^*(e^x - 1)\right)^{\frac{1}{q-1}}\rho(dx)}{\int_{-\infty}^{\infty}\left(1 + (q-1)\gamma_q^*(e^x - 1)\right)^{\frac{1}{q-1}}\rho(dx)}. \quad (7.123)$$

Notes

We remark here that the MEMM and the ESSMM are equivalent to the original probability. On the other hand the other measures (for example VOMM) are not necessarily equivalent to the original probability measure. They may be a signed martingale measure or only absolutely continuous to the original probability measure.

The basic properties of the MEMM are studied in Miyahara (1996a) [81] and Frittelli (2000) [40]. In Miyahara (1996b) [82] relations of the MEMM to the MMM are studied. For a survey of recent works on the MEMM, see Schweizer (2010) [115].

The existence problem of the MEMM for geometric Lévy processes has been studied in Miyahara (1999a) [83], Chan (1999) [18], Miyahara (2001) [86], Fujiwara and Miyahara (2003) [44], Esche and Schweizer (2005) [34], and Hubalek and Sgarra (2006) [53].

The [GLP & MEMM] pricing model was introduced in Miyahara (2001) [86], and has been studied by Miyahara and Novikov (2002) [95], Fujiwara and Miyahara (2003) [44], Miyahara and Moriwaki (2005) [92], Miyahara and Moriwaki (2006) [93], Miyahara (2006) [90], and Miyahara and Moriwaki (2009) [94]. Miyahara's works [83], [84], and [85] are related to this model.

The [GSP (geometric stable process) & MEMM] pricing model is studied again in Chapter 9.

Chapter 8

Calibration and Fitness Analysis of the [GLP & MEMM] Model

In this chapter we will study the calibration problem and the fitness analysis of [GLP & MEMM] pricing models which are formulated in Chapter 7.

8.1 The Physical World and the MEMM World

The behavior of a price process S_t is determined by the original probability (= physical probability) P, and the movement of S_t is observable in the market. These phenomena are in the real world (= physical world), and the process $S_t = S_0 e^{Z_t}$ is called a historical (or physical) price process when it is considered under the original probability P.

On the other hand the theoretical price of an option is determined through a risk-neutral measure, for example the MEMM $P^{(MEMM)}$. And the behavior of S_t under the MEMM is not observable, so this is in an ideal world (= risk-neutral world, or MEMM world). The process $S_t = S_0 e^{Z_t}$ should be called the implied price process when it is considered in the risk-neutral world. The theoretical prices of options are determined through the martingale measure in the risk-neutral world (for example the MEMM world).

8.1.1 *From the physical world to the MEMM world*

Suppose that a price process $S_t = S_0 e^{Z_t}$ is given and the generating triplet of Z_t is (σ^2, ν, b). We assume that the assumptions of Theorem 7.1 are true, and let γ^* be a solution of $\Phi(\gamma) = r$, where $\Phi(\gamma)$ is

$$\Phi(\gamma) = b + (\frac{1}{2} + \gamma)\sigma^2 + \int_{\{|x| \leq 1\}} ((e^x - 1)e^{\gamma x} - x) \, \nu(dx)$$

$$+ \int_{\{|x| > 1\}} (e^x - 1)e^{\gamma x} \, \nu(dx).$$

Then by Theorem 7.1 the MEMM $P^* = P^{(MEMM)}$ exists. (To simplify the notation we denote the MEMM by P^*.)

The generating triplet (A^*, ν^*, b^*) of Z_t under P^* is

$$A^* = \sigma^2, \tag{8.1}$$

$$\nu^*(dx) = e^{\gamma^*(e^x - 1)}\nu(dx), \tag{8.2}$$

$$b^* = b + \gamma^*\sigma^2 + \int_{\{|x| \leq 1\}} x d(\nu^* - \nu) \tag{8.3}$$

$$= b + \gamma^*\sigma^2 + \int_{\{|x| \leq 1\}} x \left(e^{\gamma^*(e^x - 1)} - 1 \right) \nu(dx). \tag{8.4}$$

This triplet determines the implied price process $(= S_t = S_0 e^{Z_t}$ under P^*) in the framework of the [GLP & MEMM] pricing model.

Remark 8.1. The condition $(C)_2$ for γ^* in Theorem 7.1 (i.e. γ^* is a solution of $\Phi(\gamma) = r$) is equivalent to the following condition (M):

$$(M) \quad b^* - r + \frac{1}{2}\sigma^2 + \int_{\{|x| \leq 1\}} (e^x - 1 - x)\nu^*(dx) + \int_{\{|x| > 1\}} (e^x - 1)\nu^*(dx) = 0. \tag{8.5}$$

It should be noted that γ^* does not appear explicitly in this formula, and that this formula is just the same with the condition that P^* is a martingale measure of the price process S_t.

8.1.2   *From the MEMM world to the physical world*

Let us study the inverse problem of the subject discussed above. Suppose that an implied price process (i.e. the price process $S_t = S_0 e^{Z_t}$ under P^*) is obtained and the generating triplet (σ^2, ν^*, b^*) of Z_t under P^* is given. Since we assume that P^* is the martingale measure, the martingale condition (M) is satisfied.

We try to construct a probability $\tilde{P}$ such that under $\tilde{P}$ the price process $S_t = S_0 e^{Z_t}$ is a geometric Lévy process and that the MEMM of $S_t = S_0 e^{Z_t}$ transformed from $\tilde{P}$ is P^*.

Let γ^* be any real number (usually we suppose that $\gamma^* < 0$), and set

$$\tilde{\nu}_{\gamma^*}(x) = e^{-\gamma^*(e^x-1)}\nu^*(dx) \tag{8.6}$$

and

$$\tilde{b}_{\gamma^*} = b^* - \gamma^*\sigma^2 + \int_{\{|x|\leq 1\}} x\left(e^{-\gamma^*(e^x-1)} - 1\right)\nu^*(dx), \tag{8.7}$$

where we assume that all integrals in the above formulas are convergent. Here we assume that we could have constructed an equivalent probability measure $\tilde{P}_{\gamma^*}$ such that Z_t is a Lévy process under $\tilde{P}_{\gamma^*}$ and the generating triplet of Z_t under $\tilde{P}_{\gamma^*}$ is $(\sigma^2, \tilde{\nu}_{\gamma^*}, \tilde{b}_{\gamma^*})$.

In the above circumstances, it is easy to see that P^* is the MEMM of $S_t = S_0 e^{Z_t}$ transformed from $\tilde{P}_{\gamma^*}$. We should remark here that, depending on a selection of γ^*, there are many geometric Lévy processes whose MEMM is just the same P^*.

8.1.3   *Calibration problem and fitness analysis*

The calibration problem of the [GLP & MEMM] pricing model is formulated as follows.

Suppose that the data of European call options are given. Then estimate a geometric Lévy process which is most fitting to the given data in the MEMM world.

We have assumed that a class of model (for example the geometric stable process model) is fixed, and the estimation is done in this class.

The fitness analysis is formulated as follows.

Suppose that the data of underlying asset prices and European call options in the market are given. Then estimate a model (namely, a class of geometric Lévy processes and a martingale measure) which is most fitting to the given data of both underlying asset prices and European call options.

8.2   Reproducibility of Volatility Smile/Smirk Property of the [GLP & MEMM] Model

In this section we see that simulation results show us that the [GLP & MEMM] model has the volatility smile/smirk property. (See [93].)

8.2.1   *Implied volatility of the model*

Suppose that an option pricing model is given. Let C_K^* be the theoretical price of a European call option with respect to the given model, and let $\sigma_K^{(im)*}$ be a solution of the following equation for σ:

$$S_0 N(d_1) - e^{-rT} K N(d_2) = C_K^*, \qquad (8.8)$$

where the variable σ is contained in $N(d_1)$ and $N(d_2)$ (see Section 1.6.2). Then $\sigma_K^{(im)*}$ is the implied volatility with respect to a given model. The implied volatility $\sigma_K^{(im)*}$ is considered to be a function of K. As a function of K, if $\sigma_K^{(im)*}$ has the volatility smile property, then the model is said to have the volatility smile property, and if $\sigma_K^{(im)*}$ has the volatility smirk property, then the model is said to have the volatility smirk property.

Remark 8.2. Equation (8.8) is obtained from (1.10) by replacing $C_K^{(m)}$ with C_K^*.

Before solving equation (8.8), we have to compute the theoretical prices of European call options for a [GLP & MEMM] model. If we know the distribution $\mu_T^*(dz)$ of Z_T under P^*, then we get $C_K^* = E_{P^*}[e^{-rT}(S_T - K)^+] = \int_K^\infty e^{-rT}(S_T - K)\mu^*(dz)$. A Lévy process is characterized by its generating

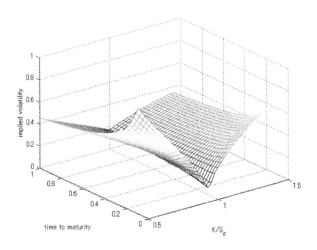

Fig. 8.1   Implied volatility surface of the [Geometric Variance Gamma Process & MEMM] model.

triplet, and the characteristic function of a Lévy process is expressed by the generating triplet. So we can assume that the characteristic function $\phi_T^*(u)$ of Z_T under P^* is given, and the distribution $\mu_T^*(dz)$ of Z_T under P^* is obtained from $\phi_T^*(u)$ by the inverse Fourier transformation. This is a method of computing the theoretical prices of European call options.

8.2.2 *[Geometric Variance Gamma Process & MEMM] model*

The first example of the [GLP & MEMM] model is the [Geometric Variance Gamma Process & MEMM] model.

From Figure 8.1, we can see that the [Geometric VG Process & MEMM] model could reproduce the volatility smile property.

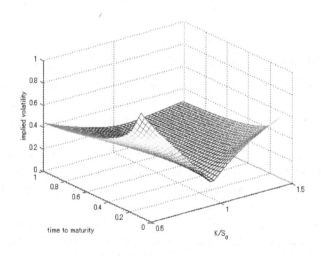

Fig. 8.2 Implied volatility surface of the [Geometric CGMY Process & MEMM] model, (a) volatility smile.

8.2.3 *[Geometric CGMY Process & MEMM] model*

The next example is the [Geometric CGMY Process & MEMM] model.

From Figures 8.2 and 8.3, we can see that the [Geometric CGMY Process & MEMM] model could reproduce the volatility smile and volatility smirk properties.

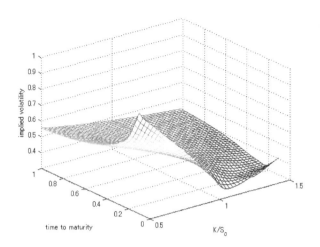

Fig. 8.3   Implied volatility surface of the [Geometric CGMY Process & MEMM] model, (b) volatility smirk.

8.2.4   *[Geometric Stable Process & MEMM] model*

The next example is the [Geometric Stable Process & MEMM] model.

From Figures 8.4 and 8.5, we can see that the [Geometric Stable Process & MEMM] model could reproduce the volatility smile and volatility smirk properties.

Moreover, changing the parameters of this model, we can see that the [Geometric Stable Process & MEMM] model could reproduce many kinds of volatility smile and volatility smirk surfaces. (See [93] for details.) Summing up the results of [93], we obtain Figure 8.6.

Figures 8.4, 8.5, and 8.6 show us that the [Geometric Stable Process & MEMM] model has the reproducibility of many types of volatility smile and smirk properties. We remark here that the strong smirk appears when the strong fat tail exists on the right-hand side.

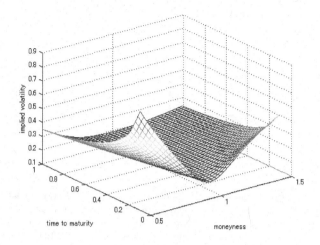

Fig. 8.4   Implied volatility surface of the [Geometric Stable Process & MEMM] model, (a) volatility smile.

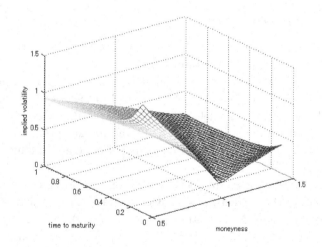

Fig. 8.5   Implied volatility surface of the [Geometric Stable Process & MEMM] model, (b) volatility smirk.

8.3   Calibration of [GLP & MEMM] Pricing Model

In this section we will study the calibration problem.

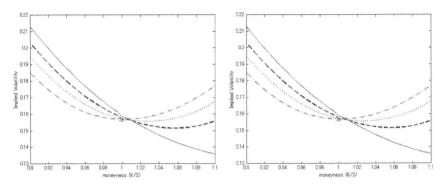

Fig. 8.6 Volatility smile/smirk properties of the [G-Stable & MEMM] model. $b = 0$, $\theta = 1.7$ and $r_d = r_f = 0$, $S = 1$, $T = 0.25$. Left figure : dash-dot line ($c_1 = 0.005, c_2 = 0.005$), dotted line ($c_1 = 0.0065, c_2 = 0.0035$), dashed line ($c_1 = 0.008, c_2 = 0.002$), solid line ($c_1 = 0.01, c_2 = 1.0e - 11$). Right figure : dash-dot line ($c_1 = 1.0e - 11, c_2 = 0.01$), dotted line ($c_1 = 0.002, c_2 = 0.008$), dashed line ($c_1 = 0.0035, c_2 = 0.0065$), solid line ($c_1 = 0.005, c_2 = 0.005$).

8.3.1 *Pricing error*

Sometimes we estimate an option-pricing model from the data of market prices of options. This is the calibration.

Suppose that we have selected a class of [GLP & MEMM] model. We denote by $C^*(K, T)$ the theoretical prices of European call options based on the selected class of model, where K is the strike price and T is the maturity. And we denote by $C^{(m)}(K, T)$ the market prices of European call options. The difference between the model price and the market price is $|C^*(K, T) - C^{(m)}(K, T)|$.

We can measure the error between the option prices of a model and option prices in the market in two ways. The first is the root mean square error (RMSE). When N theoretical prices $C^*(K_i, T_i), i = 1, 2, \ldots, N\}$ and N market data $\{C^{(m)}(K_i, T_i), i = 1, 2, \ldots, N\}$ are given, then the RMSE is defined by

$$\text{RMSE} = \sqrt{\frac{1}{N} \sum_{i=1}^{N} \left| C^*(K_i, T_i) - C^{(m)}(K_i, T_i) \right|^2}. \qquad (8.9)$$

The second one is the average relative percentage error (ARPE) which is given by

$$\text{ARPE} = \frac{1}{N} \sum_{i=1}^{N} \frac{\left| C^*(K_i, T_i) - C^{(m)}(K_i, T_i) \right|}{C^{(m)}(K_i, T_i)}. \tag{8.10}$$

8.3.2 Minimization problem

Suppose that we have selected a class of [GLP & MEMM] model (for example the geometric stable process model). Then the root mean square error (RMSE) is a candidate for the measurement of the pricing error of a model:

$$\text{RMSE}_\theta = \sqrt{\frac{1}{N} \sum_{i=1}^{N} \left| C_\theta^*(K_i, T_i) - C^{(m)}(K_i, T_i) \right|^2} \tag{8.11}$$

where θ is a parameter of the model. The problem to solve is the following minimization problem:

$$\min_\theta \text{RMSE}_\theta = \min_\theta \sqrt{\frac{1}{N} \sum_{i=1}^{N} \left| C_\theta^*(K_i, T_i) - C^{(m)}(K_i, T_i) \right|^2}. \tag{8.12}$$

Assume that this minimization problem has a solution, set $\text{RMSE}^{(cal)}$ as the minimum value, and denote the minimizing parameter by $\theta^{(cal)}$. ($^{(cal)}$ means "calibration".)

We can also adopt average relative percentage error (ARPE) as a measure of the error of the model with parameter θ:

$$\text{ARPE}_\theta = \frac{1}{N} \sum_{i=1}^{N} \frac{\left| C_\theta^*(K_i, T_i) - C^{(m)}(K_i, T_i) \right|}{C^{(m)}(K_i, T_i)}. \tag{8.13}$$

Then the minimization problem is

$$\min_\theta \text{ARPE}_\theta = \min_\theta \frac{1}{N} \sum_{i=1}^{N} \frac{\left| C_\theta^*(K_i, T_i) - C^{(m)}(K_i, T_i) \right|}{C^{(m)}(K_i, T_i)}. \tag{8.14}$$

Assuming that this minimization problem has a solution, we denote the minimum value by $\text{ARPE}^{(cal)}$ and the minimizing parameter by $\theta^{(cal)}$.

8.3.3 Procedure of calibration

The class of geometric Lévy processes is too large to carry on the calibration in this class. So we have to restrict the class of Lévy processes to, for example, the variance gamma processes, CGMY processes, or stable processes, etc.

The procedure of the calibration is as follows:

1) Collect the market data $\{C^{(m)}(K_i, T_i), i = 1, 2, \ldots, N\}$.

2) Select a class of Lévy process (for example, stable process) with parameter θ.

3) Analyze the MEMM P^* and calculate the generating triplet $(\sigma^{*2}, \nu^*, b^*)$ of the selected Lévy process under P^*. (This process is carried out in the way as we have described in Section 8.1.1.)

4) Compute the theoretical option prices $\{C_\theta^*(K_i, T_i), i = 1, 2, \ldots, N\}$ corresponding to the market prices. (This is carried out in the MEMM world.)

5) Solve the minimization problem, and set the solution as $\mathrm{RMSE}^{(cal)}$ (or $\mathrm{ARPE}^{(cal)}$), and the corresponding parameter as $\theta^{(cal)}$.

6) Then the model with parameter $\theta^{(cal)}$ is the calibrated model.

7) Repeat the same procedure for different classes of pricing model. The class of model whose $\mathrm{RMSE}^{(cal)}$ or $\mathrm{ARPE}^{(cal)}$ is the minimum is the best class among the set of classes we have investigated.

The calibration problem is "the inverse problem".

Calibration error procedure:

Physical World   MEMM World

Physical probability: P

Model $Z_t^{(\theta)}$: $(\sigma_\theta^2, \nu_\theta, b_\theta) \longrightarrow \gamma_\theta^* \longrightarrow$ MEMM: P_θ^*

$\longrightarrow (\sigma_\theta^{*2}, \nu_\theta^*, b_\theta^*)$

$\longrightarrow$ Theoretical prices: $\{C_\theta^*(K, T)\}$

Given data in the Market: $\{C^{(m)}(K_i, T_i)\}$

Calibration error $= \mathrm{RMSE}^{(cal)}$ or $\mathrm{ARPE}^{(cal)}$

$$\mathrm{RMSE}^{(cal)} = \min_\theta \sqrt{\frac{1}{N} \sum_{i=1}^{N} \left| C_\theta^*(K_i, T_i) - C^{(m)}(K_i, T_i) \right|^2}, \qquad (8.15)$$

or

$$\mathrm{ARPE}^{(cal)} = \min_\theta \frac{1}{N} \sum_{i=1}^{N} \frac{\left| C_\theta^*(K_i, T_i) - C^{(m)}(K_i, T_i) \right|}{C^{(m)}(K_i, T_i)}. \qquad (8.16)$$

$\longrightarrow \qquad \theta^{(cal)} \qquad \longrightarrow \qquad$ Calibrated model.

In Chapter 9 we will see the case of the geometric stable process model.

8.4 Fitness Analysis

Suppose that both the sequential data of prices of underlying assets and the data of market prices of European call option are given. Our question is: which model is the most fitting to the given data? To answer this problem, we need the fitness analysis of each model.

8.4.1 *Procedure of fitness analysis*

The procedure of fitness analysis is as follows

1) Select a class of Lévy processes (for example, stable processes) with the parameter θ.
2) Collect the sequential data $\{S_{t_j}, j = 1, 2, \ldots, J\}$ of the prices of underlying assets and the market data $\{C^{(m)}(K_i, T_i), i = 1, 2, \ldots, N\}$ of European call options.
3) Estimate the parameter θ of the model from the data $\{S_{t_j}, j = 1, 2, \ldots, J\}$, and denote the estimated value of θ by $\hat{\theta}$. (This procedure is carried out in the physical world.)
4) Compute the theoretical option prices $\{C^*_{\hat{\theta}}(K_i, T_i), i = 1, 2, \ldots, N\}$ under the assumption that the model parameter is $\hat{\theta}$. (This is done in the MEMM world.)
5) Compute the fitness error ϵ^*, where

$$\epsilon^* = \sqrt{\frac{1}{N} \sum_{i=1}^{N} |C^*_{\hat{\theta}}(K_i, T_i) - C^{(m)}(K_i, T_i)|^2}, \tag{8.17}$$

or

$$\epsilon^* = \frac{1}{N} \sum_{i=1}^{N} \frac{\left| C^*_{\hat{\theta}}(K_i, T_i) - C^{(m)}(K_i, T_i) \right|}{C^{(m)}(K_i, T_i)}. \tag{8.18}$$

6) Repeat the same procedure for different classes of pricing model, and the class whose ϵ^* is the minimum is the best class of model.

Remark 8.3. The calibration selects a model which is the most fitting to the market data of options. On the other hand, the fitness analysis selects a model which is the most fitting to both the market data of options and the data of the price process of the underlying assets.

The fitness analysis is important in the theoretical sense, but the calibration is considered to be more important and useful to the application.

Remark 8.4. The estimation problem for the Lévy process has been studied by many researchers, and a simple method of estimation is described in Appendix A.

Fitness analysis procedure:

Physical World   MEMM World

Physical Probability: P

Given data: $\{\xi_j = S_{t_j}\}$ (time series data)

Estimated Model $Z_t^{(\hat{\theta})}$: $(\sigma_{\hat{\theta}}^2, \nu_{\hat{\theta}}, b_{\hat{\theta}})$

$$\longrightarrow \quad \hat{\gamma}^* \quad \longrightarrow \quad \text{MEMM: } P_{\hat{\theta}}^*$$

$$\longrightarrow \quad (\sigma_{\hat{\theta}}^{*2}, \nu_{\hat{\theta}}^*, b_{\hat{\theta}}^*)$$

$$\longrightarrow \text{Theoretical prices: } \{C_{\hat{\theta}}^*(K_i, T_i)\}$$

Given data in the market: $\{C^{(m)}(K_i, T_i)\}$

Fitting error:

$$\epsilon^* = \sqrt{\frac{1}{N} \sum_{i=1}^{N} |C_{\hat{\theta}}^*(K_i, T_i) - C^{(m)}(K_i, T_i)|^2}, \qquad (8.19)$$

or

$$\epsilon^* = \frac{1}{N} \sum_{i=1}^{N} \frac{\left| C_{\hat{\theta}}^*(K_i, T_i) - C^{(m)}(K_i, T_i) \right|}{C^{(m)}(K_i, T_i)}. \qquad (8.20)$$

Notes

The results of this chapter are based on Miyahara (2006) [90], Miyahara and Moriwaki (2005) [92], and Miyahara and Moriwaki (2006) [93].

Chapter 9

The [GSP & MEMM] Pricing Model

The geometric stable process model is one of the most remarkable models for the asset price process with strong fat tail property. (See [36] and [77].) In fact, many researchers have studied this model ([33], [103], and [16]), but the suitable martingale measure for this process was not clear from their research.

We have seen in Chapter 7 that, by adopting the minimal entropy martingale measure (MEMM) as the suitable martingale measure, we can construct an option-pricing model based on the stable process under a very general setting. In fact, from Theorem 7.3 we know that the MEMM exists for a wide class of stable process models, and so we can construct a general [GSP(geometric stable process) & MEMM] model.

In this chapter we see that the [GSP & MEMM] model has a very powerful reproducibility of both the volatility smile and the volatility smirk properties. Next, by applying this model to the currency options, we can ascertain that this model fits very well to the market prices of currency options. We also investigate the calibration problem of this model.

Remark 9.1. As we have already mentioned in the previous chapters, the ESSMM and the ML^qEMM do not exist in general for stable process models.

9.1 The Physical World and the MEMM World

The physical world is the probability space $(\Omega, \mathcal{F}, P)$ where the underlying process $S_t = S_0 e^{Z_t}$ is defined. The MEMM world is $(\Omega, \mathcal{F}, P^{(MEMM)})$. For simplicity we will set $P^* = P^{(MEMM)}$.

Option Pricing in Incomplete Markets

9.1.1 *From the physical world to the MEMM world (GSP case)*

In the case of the geometric stable process, the generating triplet of Z_t is $(0, \nu(dx), b)$, where $\nu(dx)$ is

$$\nu(dx) = \frac{c_1 1_{\{x<0\}}(x) + c_2 1_{\{x>0\}}(x)}{|x|^{(\alpha+1)}} dx. \tag{9.1}$$

So the parameters of the model in the physical world are (α, c_1, c_2, b), and the domain of the parameters are

$$0 < \alpha < 2, \quad c_1, c_2 \geq 0 \ (c_1 + c_2 > 0), \quad -\infty < b < \infty. \tag{9.2}$$

The parameter γ^*, which determined the MEMM, is the solution of the following equation:

$$b + \int_{-\infty}^{\infty} \left((e^x - 1)e^{\gamma^*(e^x - 1)} - x 1_{\{|x| \leq 1\}}(x) \right) \nu(dx) = r. \tag{9.3}$$

Following Theorem 7.3, if the assumption that $c_1 > 0$ and $c_2 > 0$ is satisfied, then the above equation has a negative solution γ^*, and the MEMM exists. We restrict our discussion in this chapter to the class where the assumptions $c_1 > 0$ and $c_2 > 0$ are satisfied.

When a solution γ^* exists, the generating triplet of Z_t, $(0, \nu^*(dx), b^*)$, in the MEMM world is given by

$$\nu^*(dx) = \left(\frac{c_1 1_{\{x<0\}}(x) + c_2 1_{\{x>0\}}(x)}{|x|^{(\alpha+1)}} \right) e^{\gamma^*(e^x - 1)} dx \tag{9.4}$$

$$b^* = b + \int_{-\infty}^{\infty} x 1_{\{|x| \leq 1\}}(x) d(\nu^* - \nu). \tag{9.5}$$

Set the following:

$$\alpha^* = \alpha, \quad c_1^* = c_1, \quad c_2^* = c_2. \tag{9.6}$$

Then the generating triplet $(0, \nu^*(dx), b^*)$ in the MEMM world is determined by 5 parameters $(\gamma^*, \alpha^*, c_1^*, c_2^*, b^*)$, and the Lévy measure is given by

$$\nu^*(dx) = \left(\frac{c_1^* I_{\{x<0\}}(x) + c_2^* I_{\{x>0\}}(x)}{|x|^{(\alpha^*+1)}} \right) e^{\gamma^*(e^x - 1)} dx. \tag{9.7}$$

So the parameters in the MEMM world are $(\gamma^*, \alpha^*, c_1^*, c_2^*, b^*)$ with the domain

$$-\infty < \gamma^* < \infty, \quad 0 < \alpha^* < 2, \quad c_1^* > 0, \quad c_2^* > 0, \quad -\infty < b^* < \infty. \tag{9.8}$$

It should be noticed here that these five parameters are not independent. The three parameters (α^*, c_1^*, c_2^*) may be given freely. But (γ^*, b^*) should be given such that the following martingale condition (M) is satisfied.

Condition (M):

$$b^* + \int_{\{|x|\leq 1\}} (e^x - 1 - x)\nu^*(dx) + \int_{\{|x|>1\}} (e^x - 1)\nu^*(dx) = r. \quad (9.9)$$

So independent parameters are $(\gamma^*, \alpha^*, c_1^*, c_2^*)$, and b^* is determined from the condition (M).

9.1.2 *From the MEMM world to the physical world (GSP case)*

Suppose that in the MEMM world a set of parameters $(\gamma^*, \alpha^*, c_1^*, c_2^*, b^*)$ is given, and assume that these parameters satisfy the condition (M). This means that the generating triplet of Z_t under P^* is $(0, \nu^*(dx), b^*)$, where

$$\nu^*(dx) = \left(\frac{c_1^* I_{\{x<0\}}(x) + c_2^* I_{\{x>0\}}(x)}{|x|^{(\alpha^*+1)}} \right) e^{\gamma^*(e^x-1)} dx. \quad (9.10)$$

Then the parameters $(\bar{\alpha}, \bar{c}_1, \bar{c}_2, \bar{b})$ of the corresponding stable process in the physical world are obtained as follows.

Let us set

$$\bar{\alpha} = \alpha^*, \ \ \bar{c}_1 = c_1^*, \ \ \bar{c}_2 = c_2^*, \quad (9.11)$$

and

$$\bar{b} = b^* + \int_{\{|x|\leq 1\}} x \left(e^{-\gamma^*(e^x-1)} - 1 \right) \nu^*(dx). \quad (9.12)$$

Then $(\bar{\alpha}, \bar{c}_1, \bar{c}_2, \bar{b})$ is the parameter of the corresponding stable process in the physical world, and the generating triplet of Z_t in the physical world is $(0, \bar{\nu}(dx), \bar{b})$, where $\bar{\nu}(dx)$ is

$$\bar{\nu}(dx) = e^{-\gamma^*(e^x-1)}\nu^*(dx) = \frac{\bar{c}_1 1_{\{x<0\}}(x) + \bar{c}_2 1_{\{x>0\}}(x)}{|x|^{(\bar{\alpha}+1)}} dx. \quad (9.13)$$

9.2 Calibration by the [GSP & MEMM] Pricing Model

The calibration problem was explained in Section 8.3. We will apply the ideas obtained there to the [GSP & MEMM] pricing model.

As we have seen in Section 9.1, the model can be considered in two different worlds, the physical world and the MEMM world. So we can discuss the calibration problem in each of them.

9.2.1 Calibration in the physical world

If the model is set up in the physical world, then the model parameters are (α, c_1, c_2, b). Suppose that the data of market prices of the European call options are given. Then we can apply the procedure described in Section 8.3.3, and obtain the calibrated model parameters $(\alpha^{(cal)}, c_1^{(cal)}, c_2^{(cal)}, b^{(cal)})$. From these model parameters, a generating triplet $(0, \nu^{(cal)}(dx), b^{(cal)})$ is determined by

$$\nu^{(cal)}(dx) = \frac{c_1^{(cal)} 1_{\{x<0\}}(x) + c_2^{(cal)} 1_{\{x>0\}}(x)}{|x|^{(\alpha^{(cal)}+1)}} dx, \qquad (9.14)$$

and the calibrated price process $S_t^{(cal)}$ is given by $S_0 e^{Z_t^{(cal)}}$, where $Z_t^{(cal)}$ is a Lévy process whose generating triplet is $(0, \nu^{(cal)}(dx), b^{(cal)})$.

Starting from this calibrated geometric Lévy process $S_t^{(cal)}$, and applying the procedure of Section 7.6.3, we can get $\gamma^{(cal)*}$ and construct the MEMM $P^{(cal)*}$. And then, for any option X, $E_{P^{(cal)*}}[e^{-rT}X]$ is the estimated price of X.

9.2.2 Calibration in the MEMM world

If the model is set up in the MEMM world, then the model parameters are $(\gamma^*, \alpha^*, c_1^*, c_2^*, b^*)$ and the condition (M) is satisfied. This means that a probability space $(\Omega, \mathcal{F}^*, P^*)$ and a geometric Lévy process $S_t^* = S_0 e^{Z_t^*}$ defined on $(\Omega, \mathcal{F}^*, P^*)$ are given, and the generating triplet of Z_t^* is in the form of $(0, \nu^*(dx), b^*)$, where

$$\nu^*(dx) = \left(\frac{c_1^* I_{\{x<0\}}(x) + c_2^* I_{\{x>0\}}(x)}{|x|^{(\alpha^*+1)}} \right) e^{\gamma^*(e^x - 1)} dx. \qquad (9.15)$$

The independent parameters are $(\gamma^*, \alpha^*, c_1^*, c_2^*)$ and parameter b^* is determined from the martingale condition (M). So the parameters we have to estimate are $\theta^* = (\gamma^*, \alpha^*, c_1^*, c_2^*)$.

Suppose that the data of market prices of European call options are given. Then we can apply the minimizing procedure described in Section 8.3.3 to the calibration in the MEMM world and obtain the calibrated model parameters $\theta^{*(cal)} = (\gamma^{*(cal)}, \alpha^{*(cal)}, c_1^{*(cal)}, c_2^{*(cal)})$. The calibrated value $b^{*(cal)}$ is determined from the martingale condition (M). Now, for any option X, $E_{P_{\theta^{*(cal)}}^*}[e^{-rT}X]$ is the estimated price of X.

We note here that the Lévy measure of Z_t under $P^*_{\theta^{*(cal)}}$ is

$$\nu^{*(cal)}(dx) = \left(\frac{c_1^{*(cal)} I_{\{x<0\}}(x) + c_2^{*(cal)} I_{\{x>0\}}(x)}{|x|^{(\alpha^{*(cal)}+1)}} \right) e^{\gamma^{*(cal)}(e^x - 1)} dx,$$

(9.16)

and $b^{*(cal)}$ is determined from the martingale condition (M).

9.3 Application of the Calibrated Process to Dollar-Yen Currency Options

We apply the [GSP & MEMM] pricing model to the pricing problem of currency options.

9.3.1 *Fact: the implied volatility curve of currency options*

From the data of European call options in the market, we can compute the implied volatility curve of the currency option. Figure 9.1 shows some examples of implied volatility curves.

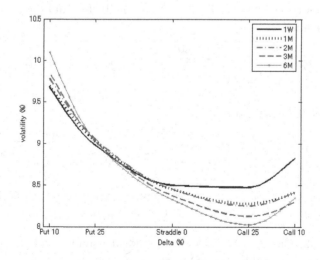

Fig. 9.1   Volatility curve of the currency options on 05/09/2005.

From this figure we know that the implied volatility curve of currency option has strong skewness. In Section 8.2 it has been shown that the [GSP & MEMM] model has the reproducibility of many types of volatility smile/smirk properties. So we can expect that the [GSP & MEMM] model will fit very well with currency options. We shall see below that this conjecture is true.

9.3.2 *In-sample analysis*

The result of the calibration is summarized in Table 9.1. The time interval is 05/09/2005–04/09/2006.

The model error at t of a model with parameter θ is measured by

$$\text{SSE}_\theta(t) := \sum_{i=1}^{n(t)} \left| C_\theta^*(t, K_i, T_i) - C^{(m)}(t, K_i, T_i) \right|^2, \quad t = 1, \ldots, N, \quad (9.17)$$

where $C_\theta^*(t, K_i, T_i)$ is the theoretical price at t of a call option based on the model with parameter θ, $C^{(m)}(t, K_i, T_i)$ is the price at t of the OTC (=over-the-counter) call option, and $n(t)$ is the number of call options at t. (In the case of our data, $n(t) = 25$.)

We calibrate the model parameters at each time t, and the calibrated parameter $\theta_t^{(cal)}$ is defined by the following condition:

$$\text{SSE}_{\theta_t^{(cal)}}(t) = \min_\theta \left\{ \text{SSE}_\theta(t) \right\}, \quad t = 1, \ldots, N. \quad (9.18)$$

We adopt two kinds of calibration error, the RMSE and the ARPE. The RMSE is defined by

$$\text{RMSE}^{(cal)} := \sqrt{ \frac{1}{\sum_{t=1}^N n(t)} \sum_{t=1}^N \sum_{i=1}^{n(t)} \left| C_{\theta_t^{(cal)}}^*(t, K_i, T_i) - C^{(m)}(t, K_i, T_i) \right|^2 }, \quad (9.19)$$

and the ARPE is defined by

$$\text{ARPE}^{(cal)} := \frac{1}{\sum_{t=1}^N n(t)} \sum_{t=1}^N \sum_{i=1}^{n(t)} \frac{\left| C_{\theta_t^{(cal)}}^*(t, K_i, T_i) - C^{(m)}(t, K_i, T_i) \right|}{C^{(m)}(t, K_i, T_i)}. \quad (9.20)$$

From Table 9.1 we know that the [GSP & MEMM] model is the most fitting model for currency options among the models listed in the table.

Remark 9.2. Carr and Wu (2003) [16] studied a similar problem using the finite moment log stable (FMLS) model.

Table 9.1   In-sample Performance

Model	Measure	Calibration Error	
		RMSE	ARPE
One-parameter models			
Black–Scholes model		0.0780	0.0595
Two-parameter models			
FMLS		0.0435	0.0549
Three-parameter models			
Scaled t-distribution	ESSMM	0.0558	0.0764
Four-parameter models			
Variance gamma process	ESSMM	0.0448	0.0534
NIG process	ESSMM	0.0544	0.0505
Merton model		0.0405	0.0642
Stable process	MEMM	0.0398	0.0392

(*Sample period*: 5/9/2005−4/9/2006.)

9.3.3   Out-of-sample analysis

The result of calibration is summarized in Table 9.2. The time interval is 05/09/2005–04/09/2006.

The model parameters are calibrated every day as before in (9.18). The theoretical option prices the next day are predicted using the calibrated model parameters; namely, it is given by $C^*_{\theta^{(cal)}_{t-1}}(t, K_i, T_i)$. The calibration error of a model is

$$
\text{RMSE}^{(cal)} = \sqrt{\frac{1}{\sum_{t=2}^N n(t)} \sum_{t=2}^N \sum_{i=1}^{n(t)} \left| C^*_{\theta^{(cal)}_{t-1}}(t, K_i, T_i) - C^{(m)}(t, K_i, T_i) \right|^2},
$$
(9.21)

and

$$
\text{ARPE}^{(cal)} = \frac{1}{\sum_{t=2}^N n(t)} \sum_{t=2}^N \sum_{i=1}^{n(t)} \frac{\left| C^*_{\theta^{(cal)}_{t-1}}(t, K_i, T_i) - C^{(m)}(t, K_i, T_i) \right|}{C^{(m)}(t, K_i, T_i)}.
$$
(9.22)

From Table 9.2 we know that the [GSP & MEMM] model is the most fitting model for currency options among the models listed in the table.

9.3.4   Volatility-based calibration

In a similar way to what we have done in above subsections, we can do a volatility-based calibration.

Table 9.2   Out-Sample Analysis

Model	Measure	Pricing Error RMSE	ARPE
One-parameter models			
Black–Scholes model		0.0798	0.0628
Two-parameter models			
FMLS		0.0467	0.0575
Three-parameter models			
Scaled t-distribution	ESSMM	0.0584	0.0779
Four-parameter models			
Variance gamma process	ESSMM	0.0480	0.0562
NIG process	ESSMM	0.0570	0.0540
Merton model		0.0440	0.0661
Stable process	MEMM	0.0432	0.0435

(*Time interval*: 05/09/2005−04/09/2006.)

Set the following:

$$\mathrm{SSE}_\theta(t) := \sum_{i=1}^{n(t)} \left| \mathrm{Vol}^{(im)}(C_\theta^*(t, K_i, T_i), t, K_i, T_i) - \sigma^{(im)}(t, K_i, T_i) \right|^2, \quad (9.23)$$

where $\mathrm{Vol}^{(im)}(C, t, K_i, T_i)$ is the implied volatility of a call-option price C, $C_\theta^*(t, K_i, T_i)$ is the theoretical price of a call option, and $\sigma^{(im)}(t, K_i, T_i)$ is the implied volatility of a call option in the market.

The estimated parameter $\hat{\theta}_t$ is defined by the following condition:

$$\mathrm{SSE}_{\hat{\theta}_t}(t) = \min_\theta \mathrm{SSE}_\theta(t). \quad (9.24)$$

In the case of in-sample analysis, the calibration error of a model is given by

$$\mathrm{RMSE}^{(cal)}$$
$$= \sqrt{\frac{1}{\sum_{t=1}^N n(t)} \sum_{t=1}^N \sum_{i=1}^{n(t)} \left| \mathrm{Vol}^{(im)}(C_{\hat{\theta}_t}^*(t, K_i, T_i), t, K_i, T_i) - \sigma^{(im)}(t, K_i, T_i) \right|^2}, \quad (9.25)$$

and

$$\mathrm{ARPE}^{(cal)}$$
$$= \frac{1}{\sum_{t=1}^N n(t)} \sum_{t=2}^N \sum_{i=1}^{n(t)} \frac{\left| \mathrm{Vol}^{(im)}(C_{\hat{\theta}_t}^*(t, K_i, T_i), t, K_i, T_i) - \sigma^{(im)}(t, K_i, T_i) \right|}{\sigma^{(im)}(t, K_i, T_i)}. \quad (9.26)$$

In the case of out-of-sample analysis, the calibration error of a model is given by

$$\text{RMSE}^{(cal)}$$

$$= \sqrt{\frac{\sum_{t=2}^{N} \sum_{i=1}^{n(t)} \left| \text{Vol}^{(im)}(C_{\hat{\theta}_{t-1}}^*(t, K_i, T_i), t, K_i, T_i) - \sigma^{(im)}(t, K_i, T_i) \right|^2}{\sum_{t=2}^{N} n(t)}},$$

$$(9.27)$$

and

$$\text{ARPE}^{(cal)}$$

$$= \frac{1}{\sum_{t=2}^{N} n(t)} \sum_{t=2}^{N} \sum_{i=1}^{n(t)} \frac{\left| \text{Vol}^{(im)}(C_{\hat{\theta}_{t-1}}^*(t, K_i, T_i), t, K_i, T_i) - \sigma^{(im)}(t, K_i, T_i) \right|}{\sigma^{(im)}(t, K_i, T_i)}.$$

$$(9.28)$$

The results are given in Table 9.3. From this table we know that the [GSP & MEMM] model is the best one.

Table 9.3   Volatility-Based Calibration

Model	Measure	In-sample		Out-of-sample	
		RMSE	ARPE	RMSE	ARPE
FMLS		0.0648	0.0544	0.0678	0.0596
Variance gamma process	ESSMM	0.0599	0.0513	0.0632	0.0561
Merton model	MCMM	0.0612	0.0527	0.0645	0.0580
Stable process	MEMM	0.0485	0.0349	0.0525	0.0428

(Sample period: 5/9/2005−4/9/2006.)

The above three tables prove that the [GSP & MEMM] model is the best among the models we have investigated. This fact suggests that the [GSP & MEMM] model should be a very effective option-pricing model.

Notes

This chapter is based on the results of Miyahara and Moriwaki (2006, 2009) [93] and [94].

Fitness analysis for the [GSP & MEMM] model is possible in the same way as described in Section 8.4.

Chapter 10

The Multi-Dimensional [GLP & MEMM] Pricing Model

In this chapter we consider the case of the d-dimensional process $(d \geq 2)$. Most of the arguments in the previous chapters can be naturally extended to multi-dimensional cases.

10.1 Multi-Dimensional Lévy Processes

We should first give the definition of a d-dimensional Lévy process.

Definition 10.1. A continuous time d-dimensional stochastic process $\{\mathbf{Z}_t = (Z_t^{(1)}, \ldots, Z_t^{(d)}), 0 \leq t \leq T\}$ defined on a probability space (Ω, F, P) is a d-dimensional Lévy process if the following conditions are satisfied.

(1) (independent increments property) For any $0 \leq t_0 < t_1 < \cdots < t_n \leq T$, $\mathbf{Z}_{t_1} - \mathbf{Z}_{t_0}, \mathbf{Z}_{t_2} - \mathbf{Z}_{t_1}, \cdots, \mathbf{Z}_{t_n} - \mathbf{Z}_{t_{n-1}}$ are independent.

(2) (stationary increments property) The distribution of $\mathbf{Z}_{t+s} - \mathbf{Z}_t$ is same for all t.

(3) $\mathbf{Z}(0) = \mathbf{0}(P - a.s.)$

(4) (stochastic continuity)

$$\forall t \geq 0, \forall \epsilon > 0, \lim_{s \to t} P(|\mathbf{Z}_s - \mathbf{Z}_t| > \epsilon) = 0. \tag{10.1}$$

(5) (cadlag) There is a set $\Omega_0 \in F$, $P(\Omega_0) = 1$ such that for $\omega \in \Omega_0$ $\mathbf{Z}_t(\omega)$ is right continuous and has the left limit.

For a d-dimensional Lévy process the results for a one-dimensional Lévy process hold true.

Theorem 10.1. (1) *Let $\{\mathbf{Z}_t\}$ be a* d-*dimensional Lévy process, and let* $\mu(d\mathbf{z})$ *be the distribution of* Z_1. *Then the distribution* $\mu(d\mathbf{z})$ *is infinitely divisible.*

(2) *Suppose that an infinitely divisible* d-*dimensional distribution* $\mu(d\mathbf{z})$ *is given. Then there exists a* d-*dimensional Lévy process* $\{\mathbf{Z}_t\}$ *such that the distribution of* $\mathbf{Z}_1$ *is* $\mu(d\mathbf{z})$. *And such a Lévy process is unique in the sense of distribution.*

Theorem 10.2. (Representation of infinitely divisible distribution). (1) *Let* $\phi(\mathbf{u})$ *be a characteristic function of an infinitely divisible* d-*dimensional distribution* $\mu(d\mathbf{z})$. *Then* $\phi(\mathbf{u})$ *is expressed as follows:*

$$\phi(\mathbf{u}) = \int_{\Re^d} e^{i\langle \mathbf{u}, \mathbf{z} \rangle} \mu(d\mathbf{z})$$

$$= \exp \left\{ -\frac{1}{2} \langle \mathbf{u}, A\mathbf{u} \rangle + \int_{|\mathbf{x}| \leq 1} (e^{i\langle \mathbf{u}, \mathbf{x} \rangle} - 1 - i\langle \mathbf{u}, \mathbf{x} \rangle) \nu(d\mathbf{x}) \right.$$

$$\left. + \int_{|\mathbf{x}| > 1} (e^{i\langle \mathbf{u}, \mathbf{x} \rangle} - 1) \nu(d\mathbf{x}) + i\langle \mathbf{b}, \mathbf{u} \rangle \right\}, \tag{10.2}$$

where $(A, \nu(d\mathbf{x}), \mathbf{b})$ *is a triplet of parameters, which is called the generating triplet, and satisfies the following [GT] condition.*

[GT] condition on the generating triplet $(A, \nu(d\mathbf{x}), \mathbf{b})$**:**
  (*i*) $A = (a_{ij})$ *is a symmetric non-negative definite* $d \times d$-*matrix*,
  (*ii*) $\nu(d\mathbf{x})$ *is a measure on* $\Re^d$, *and satisfies the following assumptions:*

$$\nu(\{\mathbf{0}\}) = 0 \quad and \quad \int_{\Re^d} (|\mathbf{x}|^2 \wedge 1) \nu(d\mathbf{x}) < \infty. \tag{10.3}$$

  (*iii*) $\mathbf{b} \in \Re^d$.

(2) *Suppose that a generating triplet* $(A, \nu(d\mathbf{x}), \mathbf{b})$, *which satisfies the [GT] condition, is given. Then the function* $\phi(\mathbf{u})$ *defined in (10.2) is a characteristic function of an infinitely divisible distribution* $\mu(d\mathbf{z})$.
(3) *The correspondence between the characteristic function* $\phi(\mathbf{u})$ *of type (10.2) and the infinitely divisible distribution* $\mu(d\mathbf{z})$ *is one-to-one.*

Since the distribution of $\mathbf{Z}_1$ is an infinitely divisible d-dimensional distribution, it is determined by a generating triplet. Let $(A, \nu(d\mathbf{x}), \mathbf{b})$ be the generating triplet of $\mathbf{Z}_1$. Then the characteristic function of $\mathbf{Z}_1$, denoted by $\phi_{\mathbf{Z}_1}(\mathbf{u})$, is expressed as (10.2). And the characteristic function of $\mathbf{Z}_t$, $\phi_{\mathbf{Z}_t}(\mathbf{u})$, is given by

$$\phi_{\mathbf{Z}_t}(\mathbf{u}) = (\phi_{\mathbf{Z}_1}(\mathbf{u}))^t. \tag{10.4}$$

Therefore a Lévy process is uniquely determined in the sense of distribution by the triplet $(A, \nu(dx), \mathbf{b})$. In this sense, the triplet $(A, \nu(dx), \mathbf{b})$ is called the generating triplet of a Lévy process $\{\mathbf{Z}_t\}$.

Remark 10.1. The generating triplet $(\sigma_j^2, \nu_j(dx_j), b_j)$ of $Z_t^{(j)}$ is

$$\sigma_j^2 = a_{jj}, \tag{10.5}$$

$$\nu_j(dx_j) = \text{j-th marginal measure of } \nu(dx), \tag{10.6}$$

$$b_j = \text{j-th component of } \mathbf{b}. \tag{10.7}$$

10.2 Multi-Dimensional Geometric Lévy Processes

A d-dimensional geometric Lévy process is defined as follows:

$$\mathbf{S}_t = (S_t^{(1)}, \ldots, S_t^{(d)}), \quad S_t^{(j)} = S_0^{(j)} e^{Z_t^{(j)}}, \quad j = 1, 2, \ldots, d \tag{10.8}$$

where $\mathbf{S}_0 = (S_0^{(1)}, \ldots, S_0^{(d)})$ is a constant vector (=starting point) and $\mathbf{Z}_t = (Z_t^{(1)}, \ldots, Z_t^{(d)})$ is a d-dimensional Lévy process. By the use of Itô's formula, this process has the following expression of stochastic differential equations

$$dS_t^{(j)} = S_{t-}^{(j)} d\hat{Z}_t^{(j)}, \quad j = 1, 2, \ldots, d, \tag{10.9}$$

where $\hat{\mathbf{Z}}_t = (\hat{Z}_t^{(1)}, \ldots, \hat{Z}_t^{(d)})$ is a d-dimensional Lévy process which corresponds to $\mathbf{Z}_t$. Let $(A, \nu(dx), \mathbf{b})$ be the generating triplet of $\mathbf{Z}_t = (Z_t^{(1)}, \ldots, Z_t^{(d)})$. Then the generating triplet of $\hat{\mathbf{Z}}_t$, $(\hat{A}, \hat{\nu}(dx), \hat{\mathbf{b}})$, is given by the following formula:

$$\hat{A} = A, \tag{10.10}$$

$$\hat{\nu}(dx) = \nu \circ J^{-1}(dx), \tag{10.11}$$

$$\hat{\mathbf{b}} = \mathbf{b} + \frac{1}{2}\mathbf{a} + \int_{\mathbf{B}} (J(\mathbf{x}) - \mathbf{x})\nu(dx)$$
$$+ \int_{\mathbf{D}_1} J(\mathbf{x})\nu(\mathbf{x}) - \int_{\mathbf{D}_2} J(\mathbf{x})\nu(\mathbf{x}), \tag{10.12}$$

where

$$J(\mathbf{x}) = (e^{x_1} - 1, \ldots, e^{x_d} - 1), \tag{10.13}$$

$$\mathbf{a} = (a_{11}, \ldots, a_{dd}), \tag{10.14}$$

and

$$\mathbf{B} = \{\mathbf{x} \in \Re^d; |\mathbf{x}| \leq 1\}, \tag{10.15}$$

$$\mathbf{D}_1 = \{\mathbf{x} \in \Re^d; |J(\mathbf{x})| \leq 1, |\mathbf{x}| > 1\}, \tag{10.16}$$

$$\mathbf{D}_2 = \{\mathbf{x} \in \Re^d; |J(\mathbf{x})| > 1, |\mathbf{x}| \leq 1\}. \tag{10.17}$$

Remark 10.2. It holds that

$$supp\ \hat{\nu}(d\mathbf{x}) \subset (-1, \infty)^d. \tag{10.18}$$

Conversely, suppose that a d-dimensional Lévy process $\hat{\mathbf{Z}}_t = (\hat{Z}_t^{(1)}, \ldots, \hat{Z}_t^{(d)})$ is given, and let $(\hat{A}, \hat{\nu}(d\mathbf{x}), \hat{\mathbf{b}})$ be the generating triplet of $\hat{\mathbf{Z}}_t$. We assume that $supp\ \hat{\nu} \subset (-1, \infty)^d$ and that the equation (10.9) has a solution $\mathbf{S}_t = (S_t^{(1)}, \ldots, S_t^{(d)})$. Setting $Z_t^{(j)} = \log S_t^{(j)} - \log S_0^{(j)}, j = 1, 2, \ldots, d$, we obtain an expression of the form (10.8). Then the generating triplet of $\mathbf{Z}_t = (Z_t^{(1)}, \ldots, Z_t^{(d)})$, $(A, \nu(d\mathbf{x}), \mathbf{b})$, is given by

$$A = \hat{A}, \tag{10.19}$$

$$\nu(d\mathbf{x}) = \hat{\nu} \circ J(d\mathbf{x}), \tag{10.20}$$

$$\mathbf{b} = \hat{\mathbf{b}} - \frac{1}{2}\hat{\mathbf{a}} - \int_{\hat{\mathbf{B}}} (\mathbf{y} - J^{-1}(\mathbf{y}))\hat{\nu}(d\mathbf{y})$$

$$- \int_{\hat{\mathbf{D}}_1} \mathbf{y}\hat{\nu}(d\mathbf{y}) + \int_{\hat{\mathbf{D}}_2} \mathbf{y}\hat{\nu}(d\mathbf{y}), \tag{10.21}$$

where

$$J^{-1}(\mathbf{y}) = (\log(1 + y_1), \ldots, \log(1 + y_d)), \tag{10.22}$$

$$\hat{\mathbf{a}} = (\hat{a}_{11}, \ldots, \hat{a}_{dd}) \quad (= \mathbf{a}), \tag{10.23}$$

and

$$\hat{\mathbf{B}} = \{\mathbf{y} \in (-1, \infty)^d; |J^{-1}\mathbf{y}| \leq 1\}, \tag{10.24}$$

$$\hat{\mathbf{D}}_1 = \{\mathbf{y} \in (-1, \infty)^d; |\mathbf{y}| \leq 1, |J^{-1}\mathbf{y}| > 1\}, \tag{10.25}$$

$$\hat{\mathbf{D}}_2 = \{\mathbf{y} \in (-1, \infty)^d; |\mathbf{y}| > 1, |J^{-1}\mathbf{y}| \leq 1\}. \tag{10.26}$$

In the above workings, we have obtained two different expressions for a d-dimensional geometric Lévy process $\mathbf{S}_t = (S_t^1, \ldots, S_t^{(d)})$,

$$S_t^{(j)} = S_0^{(j)} e^{Z_t^{(j)}} = S_0^{(j)} \mathcal{E}(\hat{Z}^{(j)})_t, \quad j = 1, 2, \ldots, d. \tag{10.27}$$

where $\mathcal{E}(\hat{Z}^{(j)})_t$ is the Doléans-Dade exponential of $\hat{Z}_t^{(j)}$ (or it is sometimes called the stochastic exponential of $\hat{Z}_t^{(j)}$), and $S_t^{(j)}$ and $\hat{Z}_t^{(j)}$ are related by (10.9).

10.3 Esscher MM and MEMM

We can consider the Esscher-transformed equivalent martingale measures and the minimal entropy martingale measures for the d-dimensional processes in the same way as for the one-dimensional case.

10.3.1 Equivalent martingale measures

Definition 10.2. Suppose that a probability space (Ω, F, P) is given. A probability measure Q on (Ω, F) is called an equivalent martingale measure of $\mathbf{S}_t = (S_t^{(1)}, \dots, S_t^{(d)})$ if $Q \sim P$ (equivalent) and $e^{-rt}S_t^{(j)}$, $j = 1, 2, \dots, d$ is $\{\mathcal{F}_t\}$-martingale.

10.3.2 Esscher martingale measures

We have studied the Esscher-transformed martingale measures for one-dimensional case in Chapter 4. The idea described there can be applied to d-dimensional cases.

Definition 10.3. Let $\mathbf{R}_t, 0 \le t \le T$, be a d-dimensional process. Then the Esscher-transformed measure of P by a risk process $\mathbf{R}_t$ and a constant vector $\mathbf{h} = (h_1, h_2, \dots, h_d)$ is the probability measure $P_{\mathbf{R}_{[0,T]},\mathbf{h}}^{(ESS)}$, defined by

$$\frac{dP_{\mathbf{R}_{[0,T]},\mathbf{h}}^{(ESS)}}{dP}\Big|_{\mathcal{F}} = \frac{e^{\langle \mathbf{h}, R_T \rangle}}{E[e^{\langle \mathbf{h}, R_T \rangle}]}, \tag{10.28}$$

and this measure transformation is called the *Esscher transformation* by a risk process $\mathbf{R}_t$ and a constant vector $\mathbf{h}$. (Remark that $P_{\mathbf{R}_{[0,T]},\mathbf{h}}^{(ESS)} = P_{\mathbf{R}_T,\mathbf{h}}^{(ES\check{S})}$.)

Definition 10.4. In the above definition, if the constant vector $\mathbf{h}$ is chosen so that the $P_{\mathbf{R}_{[0,T]},\mathbf{h}}^{(ESS)}$ is a martingale measure of $\mathbf{S}_t$, then $P_{\mathbf{R}_{[0,T]},\mathbf{h}}^{(ESS)}$ is called the "Esscher-transformed martingale measure" of $\mathbf{S}_t$ by a risk process $\mathbf{R}_t$, and it is denoted by $P_{\mathbf{R}_{[0,T]}}^{(ESSMM)}$ or $P_{\mathbf{R}_T}^{(ESSMM)}$.

For a geometric Lévy process

$$\begin{aligned}
\mathbf{S}_t = (S_t^{(1)}, \dots, S_t^{(d)}) &= (S_0^{(1)} e^{Z_t^{(1)}}, \dots, S_0^{(d)} e^{Z_t^{(d)}}) \\
&= (S_0^{(1)} \mathcal{E}(\hat{Z}^{(1)})_t, \dots, S_0^{(d)} \mathcal{E}(\hat{Z}^{(d)})_t),
\end{aligned} \tag{10.29}$$

two kinds of risk processes are usually adopted. The first one is the compound-return process:

$$\mathbf{Z}_t = (Z_t^{(1)}, \dots, Z_t^{(d)}), \tag{10.30}$$

and the second one is the simple-return process:

$$\hat{\mathbf{Z}}_t = (\hat{Z}_t^{(1)}, \dots, \hat{Z}_t^{(d)}). \tag{10.31}$$

Remark 10.3. When the compound-return process $\mathbf{Z}_t$ is adopted as a risk process, then the corresponding Esscher-transformed martingale measure (if it exists) is denoted by $P^{(ESSMM)}$ and called the "Esscher martingale measure". When the simple-return process $\hat{\mathbf{Z}}_t$ is adopted as a risk process, then the corresponding Esscher-transformed martingale measure (if it exists) is denoted by $\hat{P}^{(ESSMM)}$. As we have seen for one-dimensional cases in Chapters 4 and 6, it can be proved that the martingale measure $\hat{P}^{(ESSMM)}$ is identified with the minimal entropy martingale measure (MEMM). Therefore this martingale measure $\hat{P}^{(ESSMM)}$ is usually denoted by $P^{(MEMM)}$ and called the "minimal entropy martingale measure".

10.3.3 *Minimal entropy martingale measures*

As we have mentioned in the previous subsection, the minimal entropy martingale measure (MEMM) is identified with the "simple-return Esscher-transformed martingale measure" $\hat{P}^{(ESSMM)}$.

The MEMM was originally defined as one of the minimal distance martingale measures for the relative entropy distance function. (See Chapter 6.)

The existence theorem of the MEMM for d-dimensional geometric Lévy processes is obtained in [43], which is a natural extension of the existence theorem from Chapters 6 and 7.

Theorem 10.3. *Let* $\mathbf{S}_t$ *be a* d-*dimensional geometric Lévy process given by*

$$\mathbf{S}_t = (S_t^{(1)}, \ldots, S_t^{(d)}), \quad S_t^{(j)} = S_0^{(j)} e^{Z_t^{(j)}}, \quad j = 1, 2, \ldots, d, \qquad (10.32)$$

and assume that the following condition $(\mathbf{C}^{(d)})$ *is satisfied.*

$(\mathbf{C}^{(d)})$: *A constant vector* $\gamma^* = (\gamma_1^*, \ldots, \gamma_d^*)$ *exists which satisfies the following conditions:*

$$\int_{\{|\mathbf{x}|>1\}} e^{x_j} e^{\langle \gamma^*, J(\mathbf{x}) \rangle} \nu(d\mathbf{x}) < \infty, \quad j = 1, \ldots, d, \qquad (10.33)$$

and

$$b_j + \frac{1}{2} a_{jj} + (\gamma^* \mathbf{A})_j + \int_{\{|\mathbf{x}| \le 1\}} \left((e^{x_j} - 1) e^{\langle \gamma^*, J(\mathbf{x}) \rangle} - x_j \right) \nu(d\mathbf{x})$$

$$+ \int_{\{|\mathbf{x}|>1\}} (e^{x_j} - 1) e^{\langle \gamma^*, J(\mathbf{x}) \rangle} \nu(d\mathbf{x}) = r, \quad j = 1, \ldots, d. \qquad (10.34)$$

Then the MEMM $P^{(MEMM)}$ *exists and it holds that:*
(i) $P^{(MEMM)}$ *is identified with the* $\hat{P}^{(ESSMM)}$ (*Esscher-transformed martingale measure by the simple risk process* $\hat{\mathbf{Z}}_t$),

(ii) $\mathbf{Z}_t$ is a Lévy process also w.r.t. $P^{(MEMM)}$, and the generating triplet $(A^*, \nu^*, \mathbf{b}^*)$ of $\mathbf{Z}_t$ under $P^{(MEMM)}$ is

$$A^* = A, \tag{10.35}$$

$$\nu^*(d\mathbf{x}) = e^{\langle \gamma^*, J(\mathbf{x}) \rangle} \nu(d\mathbf{x}), \tag{10.36}$$

$$\mathbf{b}^* = \mathbf{b} + \gamma^* A + \int_{\Re^d} \mathbf{x} I_{\{|\mathbf{x}| \leq 1\}} d(\nu^* - \nu), \tag{10.37}$$

(iii) *The minimal entropy distance* $H(P^{(MEMM)}|P)$ *is*

$$H(P^{(MEMM)}|P) = T \left\{ \frac{1}{2} \langle \gamma^* A, \gamma^* \rangle \right.$$
$$\left. + \int_{\Re^d} \left(\langle \gamma, J(\mathbf{x}) \rangle e^{\langle \gamma, J(\mathbf{x}) \rangle} - e^{\langle \gamma, J(\mathbf{x}) \rangle} + 1 \right) \nu(dx) \right\}.$$
$$\tag{10.38}$$

10.3.4 *[GLP & MEMM] pricing model*

Adopting a d-dimensional geometric Lévy process as a underlying asset process and the MEMM as a suitable martingale measure, we obtain the d-dimensional [GLP & MEMM] pricing model. Using this model, we can continue the discussions on option pricing in incomplete markets for multi-dimensional cases as we have done in Chapters 7, 8, and 9 for one-dimensional cases.

10.4   Application to Portfolio Evaluation

10.4.1   *Geometric Lévy process portfolio model*

Suppose that a non-risky asset price process

$$S_t^{(0)} = S_0^{(0)} e^{rt} \tag{10.39}$$

and a d-dimensional stock price process

$$\mathbf{S}_t = (S_t^{(1)}, \ldots, S_t^{(d)}), \tag{10.40}$$

$$S_t^{(j)} = S_0^{(j)} e^{Z_t^{(j)}} \quad j = 1, 2, \ldots, d, \tag{10.41}$$

are given, where $r \ (\geq 0)$ is a constant and $\mathbf{Z}_t = (Z_t^{(1)}, \ldots, Z_t^{(d)})$ is a d-dimensional Lévy process. As we have seen in Section 10.2, those processes satisfy the following stochastic differential equations:

$$\begin{cases} dS_t^{(0)} = S_t^{(0)} r dt \\ dS_t^{(j)} = S_{t-}^{(j)} d\hat{Z}_t^{(j)} \quad j = 1, 2, \ldots, d, \end{cases} \tag{10.42}$$

where $\hat{\mathbf{Z}}_t = (\hat{Z}_t^{(1)}, \ldots, \hat{Z}_t^{(d)})$ is a d-dimensional Lévy process, which is determined corresponding to $\mathbf{Z}_t$. Since $S_t^{(j)} > 0$, the condition *supp* $\hat{\nu}(d\mathbf{x}) \subset$ $(-1, \infty)^d$ is satisfied. From the above equation it follows that $S_t^{(j)}$, $j = 1, 2, \ldots, d$, are expressed as

$$S_t^{(j)} = S_0^{(j)} e^{Z_t^{(j)}} = S_0^{(j)} \mathcal{E}(\hat{Z}^{(j)})_t \quad j = 1, 2, \ldots, d, \qquad (10.43)$$

where $\mathcal{E}(\hat{Z}^{(j)})_t$ is the Doléans-Dade exponential of $\hat{Z}_j$.

Let $\theta_t = (\theta_t^{(0)}, \theta_t^{(1)}, \ldots, \theta_t^{(d)})$ be an investment strategy and let $V_t^{(\theta)} = \sum_{j=0}^d \theta_{t-}^{(j)} S_t^{(j)}$ be the corresponding value process. We assume that every investment strategy satisfies the self-financing assumption, i.e. that we have:

$$V_t^{(\theta)} = V_0 + \sum_{j=0}^d \int_0^t \theta_{t-}^{(j)} dS_t^{(j)}. \qquad (10.44)$$

Set the following:

$$w_t^{(j)} = \frac{\theta_{t-}^{(j)} S_t^{(j)}}{V_t^{(\theta)}}, \quad \left(\theta_{t-}^{(j)} = \frac{w_t^{(j)} V_t^{(\theta)}}{S_t^{(j)}} \right). \qquad (10.45)$$

Then $\{w_t^{(0)}, w_t^{(1)}, \ldots, w_t^{(d)}\}$, $\sum_{j=0}^d w_t^{(j)} = 1$, is a portfolio and $V_t^{(\theta)}$ satisfies the following equation:

$$dV_t^{(\theta)} = \sum_{j=0}^d \theta_{t-}^{(j)} dS_t^{(j)} = \sum_{j=0}^d \frac{w_{t-}^{(j)} V_{t-}^{(\theta)}}{S_{t-}^{(j)}} dS_t^{(j)} = V_{t-}^{(\theta)} \sum_{j=0}^d w_{t-}^{(j)} d\hat{Z}_t^{(j)}. \qquad (10.46)$$

10.4.2 *Risk-sensitive value measure*

We adopt the risk-sensitive value measure for the optimality criterion. (See [91] for details.)

Definition 10.5. (Risk-sensitive value measure). Let α be a real number, and $\mathbf{X}$ be the set of all random variables defined on (Ω, F, P). Then a function of $X \in \mathbf{X}$ defined as

$$U^{(\alpha)}(X) = \begin{cases} -\frac{1}{\alpha} \log E\left[e^{-\alpha X}\right], & (\alpha \neq 0), \\ E[X], & (\alpha = 0). \end{cases} \qquad (10.47)$$

is called the "risk-sensitive value measure with risk-sensitivity parameter α".

Definition 10.6. (Risk-sensitive dynamic value measure). A function $U_t^{(\alpha)}(X)$ of X defined by

$$U_t^{(\alpha)}(X) = \begin{cases} -\frac{1}{\alpha} \log E \left[e^{-\alpha X} | \mathcal{F}_t \right], & 0 \le t \le T, \quad (\alpha \ne 0), \\ E[X|\mathcal{F}_t], & 0 \le t \le T, \quad (\alpha = 0) \end{cases} \tag{10.48}$$

is called the "risk-sensitive dynamic value measure with risk-sensitivity parameter α".

Remark 10.4. (1) The risk-sensitive value measure is sometimes called the "entropic value measure".

(2) The risk-sensitive dynamic value measure is sometimes called the "entropic value measure".

(3) If $\alpha > 0$ then $U^{(\alpha)}(X)$ is a risk-averse valuation of X. If $\alpha < 0$ then $U^{(\alpha)}(X)$ is a risk-loving valuation of X. When $E\left[e^{-\alpha X} \right] = \infty$, it holds that

$$U^{(\alpha)}(X) = -\infty \;\; if \;\; \alpha > 0, \tag{10.49}$$

$$U^{(\alpha)}(X) = \infty \;\; if \;\; \alpha < 0. \tag{10.50}$$

(4) We formally obtain the following relation:

$$\lim_{\alpha \to 0} U^{(\alpha)}(X) = U^{(0)}(X). \tag{10.51}$$

10.4.3   *Risk-sensitive evaluation of portfolio*

Let Θ be the set of all self-finance portfolio strategies. When we adopt the risk-sensitive value measure, then the optimal portfolio problem is reduced to the following optimization problem:

$$\sup_{\theta \in \Theta} \left\{ U^{(\alpha)}(V_T^{(\theta)}) \right\} = -\frac{1}{\alpha} \log \left(\inf_{\theta \in \Theta} \left\{ E \left[e^{-\alpha V_T^{(\theta)}} \right] \right\} \right), \tag{10.52}$$

where we assume that $\alpha > 0$.

We introduce a new kind of optimal portfolio as follows.

Definition 10.7. The value

$$\sup_{\theta \in \Theta} \left\{ U^{(\alpha)}(V_T^{(\theta)}) \right\} \tag{10.53}$$

is called the "risk-sensitive value of the return optimal portfolio" based on the assets $(S_t^{(1)}, \dots, S_t^{(d)})$.

The above optimization problem (10.52) is equivalent to the following maximization problem:

$$\sup_{\theta \in \Theta} \left\{ E \left[\frac{1}{\alpha} \left(1 - e^{-\alpha V_T^{(\theta)}} \right) \right] \right\}. \tag{10.54}$$

This is the expected utility optimization problem with the following exponential utility function:

$$u_\alpha(x) = \frac{1}{\alpha} \left(1 - e^{-\alpha x} \right). \tag{10.55}$$

A powerful method of solving the expected utility optimization problem is the duality method. (See, for example [69] and [100].) We can apply this method to the above problem, and obtain the following formula:

$$\sup_{\theta \in \Theta} E \left[u_\alpha \left(V_T^{(\theta)} \right) \right] = \sup_{\theta \in \Theta} E \left[u_\alpha \left(V_0 + \sum_{j=0}^{d} \int_0^T \theta_{t-}^{(j)} dS_t^{(j)} \right) \right]$$

$$= \inf_{\lambda > 0} \left\{ \inf_{Q \in \mathcal{M}} \left\{ E \left[u_\alpha^* \left(\lambda \frac{dQ}{dP} \right) \right] \right\} + \lambda V_0 \right\}, \tag{10.56}$$

where

$$u_\alpha^*(y) = \sup_{x \in \Re} (u_\alpha(x) - xy) = \frac{1}{\alpha} \left(y \log y + 1 - y \right). \tag{10.57}$$

Using (10.57), we can obtain

$$\inf_{Q \in \mathcal{M}} \left\{ E \left[u_\alpha^* \left(\lambda \frac{dQ}{dP} \right) \right] \right\} = \inf_{Q \in \mathcal{M}} \left\{ \frac{1}{\alpha} \left(E \left[\lambda \frac{dQ}{dP} \log \left(\lambda \frac{dQ}{dP} \right) + 1 - \lambda \frac{dQ}{dP} \right] \right) \right\}$$

$$= \inf_{Q \in \mathcal{M}} \left\{ \frac{1}{\alpha} \left(\lambda H(Q|P) + \lambda \log \lambda - \lambda + 1 \right) \right\}$$

$$= \frac{1}{\alpha} \left(\lambda H(P^{(MEMM)}|P) + \lambda \log \lambda - \lambda + 1 \right). \tag{10.58}$$

From (10.56) and (10.58) we can obtain

$$\sup_{\theta \in \Theta} E \left[u_\alpha \left(V_T^{(\theta)} \right) \right] = \inf_{\lambda > 0} \left\{ \inf_{Q \in \mathcal{M}} E \left[u_\alpha^* \left(\lambda \frac{dQ}{dP} \right) \right] + \lambda V_0 \right\}$$

$$= \inf_{\lambda > 0} \left\{ \frac{1}{\alpha} \left(\lambda H(P^{(MEMM)}|P) + \lambda \log \lambda - \lambda + 1 \right) + \lambda V_0 \right\}$$

$$= \frac{1}{\alpha} \left(1 - e^{-\left(H(P^{(MEMM)}|P) + \alpha V_0 \right)} \right). \tag{10.59}$$

Combining the above results, we obtain

$$\sup_{\theta \in \Theta} \left\{ U^{(\alpha)}(V_T^{(\theta)}) \right\} = -\frac{1}{\alpha} \log \left(\inf_{\theta \in \Theta} \left\{ E \left[e^{-\alpha V_T^{(\theta)}} \right] \right\} \right)$$

$$= -\frac{1}{\alpha} \log \left(\inf_{\theta \in \Theta} \left\{ E \left[1 - \alpha u_\alpha \left(V_T^{(\theta)} \right) \right] \right\} \right)$$

$$= -\frac{1}{\alpha} \log \left(1 - \alpha \sup_{\theta \in \Theta} \left\{ E \left[u_\alpha \left(V_T^{(\theta)} \right) \right] \right\} \right)$$

$$= -\frac{1}{\alpha} \log \left(1 - \alpha \left(\frac{1}{\alpha} \left(1 - e^{-\left(H(P^{(MEMM)}|P) + \alpha V_0 \right)} \right) \right) \right)$$

$$= V_0 + \frac{1}{\alpha} H(P^{(MEMM)}|P). \tag{10.60}$$

We can summarize the above results in the following theorem.

Theorem 10.4. *Under the situation described above, "the risk-sensitive value of the return optimal portfolio" is given by*

$$\sup_{\theta \in \Theta} \left\{ U^{(\alpha)}(V_T^{(\theta)}) \right\} = V_0 + \frac{1}{\alpha} H(P^{(MEMM)}|P). \tag{10.61}$$

It should be noted that the explicit form of $H(P^{(MEMM)}|P)$ is obtained in Theorem 10.3 as follows:

$$H(P^{(MEMM)}|P)$$

$$= \left\{ \frac{1}{2} \langle \gamma^* A, \gamma^* \rangle + \int_{\Re^d} \left(\langle \gamma^*, J(\mathbf{x}) \rangle e^{\langle \gamma^*, J(\mathbf{x}) \rangle} - e^{\langle \gamma^*, J(\mathbf{x}) \rangle} + 1 \right) \nu(d\mathbf{x}) \right\} T. \tag{10.62}$$

Therefore, using this formula and theorem 10.4, we can analyze the risk-sensitive value of the optimal portfolio.

Remark 10.5. It is known that the optimal strategy is a re-balancing portfolio such that

$$\theta_t^j S_{t-}^j = \gamma^j, \quad j = 0, 1, \ldots, d. \tag{10.63}$$

(See [43].)

10.4.4 *Re-balancing portfolios*

We restrict the class of portfolios to time-independent portfolios, $(w^{(0)}, w^{(1)}, \ldots, w^{(d)})$. Then the corresponding value process $V^{(w)}$ satisfies the following equation:

$$dV_t^{(w)} = V_{t-}^{(w)} \sum_{j=0}^{d} w^{(j)} dV_t^{(w)} = V_{t-}^{(w)} d\hat{Z}_t^{(w)}, \tag{10.64}$$

where

$$\hat{Z}_t^{(w)} = \sum_{j=0}^{d} w^{(j)} \hat{Z}_t^{(j)}. \tag{10.65}$$

The $V_t^{(w)}$ is expressed as

$$V_t^{(w)} = V^{(w)}(0)\mathcal{E}(\hat{Z}^{(w)})_t = V^{(w)}(0)e^{Z_t^{(w)}}, \tag{10.66}$$

where $Z_t^{(w)}$ is a Lévy process corresponding to $\hat{Z}_t^{(w)}$.

In this case the optimization problem is given in the following form:

$$\sup_w \left\{ U^{(\alpha)}(V_T^{(w)}) \right\} = -\frac{1}{\alpha} \log \left(\inf_w \left\{ E\left[e^{-\alpha V_T^{(w)}} \right] \right\} \right). \tag{10.67}$$

Remark 10.6. From Remark 10.5 and Theorem 10.4, it follows that

$$\sup_w \left\{ U^{(\alpha)}(V_T^{(w)}) \right\} = \sup_{\theta \in \Theta} \left\{ U^{(\alpha)}(V_T^{(\theta)}) \right\} = V_0 + \frac{1}{\alpha} H(P^{(MEMM)}|P).$$
$$\tag{10.68}$$

10.5   Risk-Sensitive Evaluation of Growth Rate

We introduce the idea of the "random inner rate of return" (RIRR). The inner rate of return (IRR) is well known in the theory of corporate finance. The RIRR is a randomization of IRR.

Definition 10.8. (random inner rate of return). Let $\{V_t; 0 \le t \le T\}$ be a price process (or a value process) of an asset (or a portfolio). Under the assumption that $V_t > 0$, the $r_T(V.)$ defined by the following relation

$$e^{r_T(V.)T} = \frac{V_T}{V_0} \tag{10.69}$$

is well defined and called the RIRR of V_t.

From the definition of $r_T(V.)$, we obtain

$$r_T(V.) = \frac{1}{T}(\log V_T - \log V_0). \tag{10.70}$$

Remark 10.7. A sufficient condition for the assumption of $V_t^{(\theta)} > 0$ is

$$supp\ \hat{\nu}(d\mathbf{x}) \subseteq (-1,1)^d. \tag{10.71}$$

This fact follows from (10.46).

Remark 10.8. The RIRR is just the same as the "growth rate" of V_t. The portfolio which maximizes $E[\log V_T^{(\theta)}]$ is called the "growth optimal portfolio" (GOP).

10.5.1 *Risk-sensitive evaluation of RIRR*

Let us assume that $r_T(V^{(\theta)}_\cdot)$ is well defined for all portfolios where $\theta = (\theta_t)$, and investigate the following optimization problem:

$$\sup_{\theta \in \Theta} \left\{ U^{(\alpha)}(r_T(V^{(\theta)}_\cdot)) \right\} = -\frac{1}{\alpha} \log \left(\inf_{\theta \in \Theta} \left\{ E \left[e^{-\alpha r_T(V^{(\theta)}_\cdot)} \right] \right\} \right), \quad (10.72)$$

where we assume that $\alpha > 0$.

Definition 10.9. The value

$$\sup_{\theta \in \Theta} \left\{ U^{(\alpha)}(r_T(V^{(\theta)}_\cdot)) \right\} \quad (10.73)$$

is called the "risk-sensitive value of the growth optimal portfolio" based on the assets $(S^{(1)}_t, \ldots, S^{(d)}_t)$.

The value of (10.73) is calculated as follows:

$$\sup_{\theta \in \Theta} \left\{ U^{(\alpha)}(r_T(V^{(\theta)}_\cdot)) \right\} = -\frac{1}{\alpha} \log \left(\inf_{\theta \in \Theta} \left\{ E \left[e^{-\alpha r_T(V^{(\theta)}_\cdot)} \right] \right\} \right)$$

$$= -\frac{1}{\alpha} \log \left(\inf_{\theta \in \Theta} \left\{ E \left[e^{-\alpha \frac{1}{T}(\log V^{(\theta)}_T - \log V_0)} \right] \right\} \right)$$

$$= -\frac{1}{\alpha} \log \left(V_0^{\frac{\alpha}{T}} \inf_{\theta \in \Theta} \left\{ E \left[\left(V^{(\theta)}_T \right)^{-\frac{\alpha}{T}} \right] \right\} \right)$$

$$= -\frac{1}{\alpha} \log \left(\inf_{\theta \in \Theta} \left\{ E \left[\left(V^{(\theta)}_T \right)^{-\frac{\alpha}{T}} \right] \right\} \right) - \frac{1}{T} \log (V_0).$$
$$(10.74)$$

This optimization problem is equivalent to the following maximization problem:

$$\sup_{\theta \in \Theta} \left\{ E \left[- \left(V^{(\theta)}_T \right)^{-\frac{\alpha}{T}} \right] \right\}. \quad (10.75)$$

Namely this problem is equivalent to the expected utility optimization problem with respect to the power utility function

$$u_p(x) = \frac{1}{p} x^p, \quad p = -\frac{\alpha}{T} < 0. \quad (10.76)$$

We can apply the duality method to this utility optimization problem, and so obtain:

$$\sup_{\theta \in \Theta} E \left[u_p \left(V^{(\theta)}_T \right) \right] = \sup_{\theta \in \Theta} E \left[u_p \left(V_0 + \sum_{j=0}^{d} \int_0^T \theta^{(j)}_{t-} dS^{(j)}_t \right) \right]$$

$$= \inf_{\lambda > 0} \left\{ \inf_{Q \in \mathcal{M}} \left\{ E \left[u_p^* \left(\lambda \frac{dQ}{dP} \right) \right] \right\} + \lambda V_0 \right\}. \quad (10.77)$$

In this formula, $u_p^*(y)$ is the dual function of $u_p(x)$ and it is given by

$$u_p^*(y) = \sup_{x \in \Re}(u_p(x) - xy) = -\frac{1}{q}y^q, \qquad (10.78)$$

where

$$q = \frac{p}{p-1} = \frac{\alpha}{\alpha + T}. \qquad (10.79)$$

Note that, by the assumption that $\alpha > 0$, it holds that $0 < q < 1$.
Using (10.78), we obtain:

$$\begin{aligned}
&\inf_{Q \in \mathcal{M}} \left\{ E\left[u_p^*\left(\lambda \frac{dQ}{dP} \right) \right] \right\} \\
&= \inf_{Q \in \mathcal{M}} \left\{ E\left[-\frac{1}{q} \left(\lambda \frac{dQ}{dP} \right)^q \right] \right\} \\
&= \lambda^q \inf_{Q \in \mathcal{M}} \left\{ E\left[-\frac{1}{q} \left(\frac{dQ}{dP} \right)^q \right] \right\} \\
&= \lambda^q E\left[-\frac{1}{q} \left(\frac{dP^{(ML^q EMM)}}{dP} \right)^q \right], \qquad (10.80)
\end{aligned}$$

where $P^{(ML^q EMM)}$ is the minimal L^q equivalent martingale measure defined in Chapter 6. (Our discussions in Chapter 6 are one-dimensional cases; for the above result, we must investigate multi-dimensional cases.)
From (10.77) and (10.80) we obtain the following result:

$$\begin{aligned}
&\sup_{\theta \in \Theta} E\left[u_p\left(V_T^{(\theta)} \right) \right] \\
&= \inf_{\lambda > 0} \left\{ \inf_{Q \in \mathcal{M}} \left\{ E\left[u_p^*\left(\lambda \frac{dQ}{dP} \right) \right] \right\} + \lambda V_0 \right\} \\
&= \inf_{\lambda > 0} \left\{ \lambda^q E\left[-\frac{1}{q} \left(\frac{dP^{(ML^q EMM)}}{dP} \right)^q \right] + \lambda V_0 \right\} \\
&= \frac{q-1}{q} V_0^{\frac{q}{q-1}} E\left[\left(\frac{dP^{(ML^q EMM)}}{dP} \right)^q \right]^{\frac{1}{1-q}}. \qquad (10.81)
\end{aligned}$$

Combining all the above results, we obtain the following formula:

$$\sup_{\theta \in \Theta} \left\{ U^{(\alpha)}(r_T(V_{\cdot}^{(\theta)})) \right\}$$

$$= -\frac{1}{\alpha} \log \left(\inf_{\theta \in \Theta} \left\{ E \left[\left(V_T^{(\theta)} \right)^{-\frac{\alpha}{T}} \right] \right\} \right) - \frac{1}{T} \log (V_0)$$

$$= -\frac{1}{\alpha} \log \left(p \sup_{\theta \in \Theta} \left\{ E \left[u_p \left(V_T^{(\theta)} \right) \right] \right\} \right) - \frac{1}{T} \log (V_0)$$

$$= -\frac{1}{\alpha} \log \left(p \left(\frac{q-1}{q} V_0^{\frac{q}{q-1}} E \left[\left(\frac{dP^{(ML^q EMM)}}{dP} \right)^q \right]^{\frac{1}{1-q}} \right) \right) - \frac{1}{T} \log (V_0)$$

$$= -\frac{1}{\alpha(1-q)} \log \left(E \left[\left(\frac{dP^{(ML^q EMM)}}{dP} \right)^q \right] \right)$$

$$= -\frac{1}{\alpha} \left(\frac{\alpha+T}{T} \right) \log \left(E \left[\left(\frac{dP^{(ML^q EMM)}}{dP} \right)^q \right] \right). \qquad (10.82)$$

We can summarize the above results in the following theorem.

Theorem 10.5. *Under the situation described above, the risk-sensitive value of the growth optimal portfolio is equal to*

$$\sup_{\theta \in \Theta} \left\{ U^{(\alpha)}(r_T(V_{\cdot}^{(\theta)})) \right\} = -\frac{1}{\alpha} \left(\frac{\alpha+T}{T} \right) \log \left(E \left[\left(\frac{dP^{(ML^q EMM)}}{dP} \right)^q \right] \right),$$

$$\qquad (10.83)$$

where $q = \frac{\alpha}{\alpha+T}$.

Remark 10.9. In Section 6.3 the explicit form of $\frac{dP^{(ML^q EMM)}}{dP}$ is given in one-dimensional case. We can calculate the explicit form of $\frac{dP^{(ML^q EMM)}}{dP}$ in multi-dimensional cases with the same method we used in Section 6.3. In [57] multi-dimensional cases are discussed.

Remark 10.10. It is known that the optimal strategy of the above problem is a re-balancing portfolio. (See [98].)

10.5.2 Risk-sensitive evaluation of re-balancing portfolios

We start from the following equation (see (10.42)):

$$\begin{cases} dS_t^{(0)} = S_t^{(0)} r \, dt \\ dS_t^{(j)} = S_{t-}^{(j)} d\hat{Z}_t^{(j)} \quad j = 1, 2, \ldots, d. \end{cases} \qquad (10.84)$$

Assume that the process $\hat{\mathbf{Z}}_t = (\hat{Z}_t^{(1)}, \ldots, \hat{Z}_t^{(d)})$ is given and that the generating triplet of $\hat{\mathbf{Z}}_t$ is $(\hat{A}, \hat{\nu}(d\mathbf{x}), \hat{\mathbf{b}}_0)_0$. Then the characteristic function of $\hat{\mathbf{Z}}_t$ is

$$\phi_{\hat{Z}}(\mathbf{u}) = \mathrm{E}\left[e^{i\langle \mathbf{u}, \hat{\mathbf{Z}}(1)\rangle}\right]$$

$$= \exp\left\{-\frac{1}{2}\langle \mathbf{u}, \hat{A}\mathbf{u}\rangle + \int_{\Re^d}(e^{i\langle \mathbf{u},\mathbf{x}\rangle} - 1)\hat{\nu}(d\mathbf{x}) + i\langle \hat{\mathbf{b}}_0, \mathbf{u}\rangle\right\}. \quad (10.85)$$

We restrict the class of portfolios to time-independent portfolios, and denote a time-independent portfolio by $(w^{(0)}, w^{(1)}, \ldots, w^{(d)})$. Then the corresponding value process $V^{(w)}$ satisfies the following equation:

$$dV_t^{(w)} = V_{t-}^{(w)} \sum_{j=0}^d w^{(j)} dV_t^{(w)} = V_{t-}^{(w)} d\hat{Z}_t^{(w)}, \quad (10.86)$$

where

$$\hat{Z}_t^{(w)} = \sum_{j=0}^d w^{(j)} \hat{Z}_t^{(j)}. \quad (10.87)$$

The $V_t^{(w)}$ is expressed as

$$V_t^{(w)} = V^{(w)}(0)\mathcal{E}(\hat{Z}^{(w)})_t = V^{(w)}(0)e^{Z_t^{(w)}}, \quad (10.88)$$

where $Z_t^{(w)}$ is a Lévy process corresponding to $\hat{Z}_t^{(w)}$.

We consider such cases where the assumption that $V_t^{(w)} > 0$ for all w is satisfied. For example, the case where $\hat{\nu}(d\mathbf{X}) \subseteq (-1, 1)^d$ is one of such cases. Then our optimization problem is formulated as follows:

$$\sup_w \left\{ U^{(\alpha)}(r_T(V^{(w)})) \right\} = \sup_w \left\{ -\frac{1}{\alpha} \log E\left[e^{-\alpha r_T(V^{(w)})}\right] \right\}$$

$$= \sup_w \left\{ -\frac{1}{\alpha} \log E\left[e^{-\frac{\alpha}{T}(\log V_T^{(w)} - \log V_0^{(w)})}\right] \right\}$$

$$= \sup_w \left\{ -\frac{1}{\alpha} \log E\left[e^{-\frac{\alpha}{T} Z_T^{(w)}}\right] \right\}$$

$$= \sup_w \left\{ -\frac{1}{\alpha} \log \left(\phi_{Z^{(w)}}(i\frac{\alpha}{T})\right)^T \right\}$$

$$= \sup_w \left\{ -\frac{T}{\alpha} \log \left(\phi_{Z^{(w)}}(i\frac{\alpha}{T})\right) \right\}, \quad (10.89)$$

where $\phi_{Z^{(w)}}(u)$ is the characteristic function of $Z_1^{(w)}$.

We next calculate the generating triplet of $\hat{Z}_t^{(w)}$. From the following equality:

$$\hat{Z}_t^{(w)} = \sum_{j=0}^{d} w^{(j)} \hat{Z}_t^{(j)} = w^{(0)} rt + \langle \mathbf{w}, \hat{\mathbf{Z}}_t \rangle, \quad \mathbf{w} = (w^{(1)}, \ldots, w^{(d)}), \quad (10.90)$$

we obtain

$$\phi_{\hat{Z}^{(w)}}(u) = \mathrm{E}\left[e^{iu\hat{Z}_1^{(w)}} \right] = e^{irw^{(0)}u} \phi_{\hat{Z}}(u\mathbf{w})$$

$$= \exp\left\{ -\frac{1}{2}\langle \mathbf{w}, \hat{A}\mathbf{w} \rangle u^2 + \int_{\Re^d} (e^{i\langle \mathbf{w},\mathbf{x}\rangle u} - 1)\hat{\nu}(d\mathbf{x}) + i(rw^{(0)} + \langle \hat{\mathbf{b}}_0, \mathbf{w} \rangle)u \right\}.$$

$$(10.91)$$

So the generating triplet of $\hat{Z}_t^{(w)}$, denoted by $((\hat{\sigma}^{(w)})^2, \hat{\nu}^{(w)}(dx), \hat{b}_0^{(w)})_0$, is

$$(\hat{\sigma}^{(w)})^2 = \langle \mathbf{w}, \hat{A}\mathbf{w} \rangle, \quad (10.92)$$

$$\hat{\nu}^{(w)}(dx) = \left(\hat{\nu} \circ K^{-1} \right)(dx), \quad (10.93)$$

$$\hat{b}_0^{(w)} = rw^{(0)} + \langle \hat{\mathbf{b}}_0, \mathbf{w} \rangle, \quad (10.94)$$

where

$$K(\mathbf{x}) = \langle \mathbf{w}, \mathbf{x} \rangle : \Re^d \to \Re^1. \quad (10.95)$$

And the generating triplet, denoted by $((\sigma^{(w)})^2, \nu^{(w)}(dx), b_0^{(w)})_0$, of $Z_t^{(w)}$, which corresponds to $\hat{Z}_t^{(w)}$, is

$$(\sigma^{(w)})^2 = (\hat{\sigma}^{(w)})^2 = \langle \mathbf{w}, \hat{A}\mathbf{w} \rangle \quad (10.96)$$

$$\nu^{(w)}(dx) = \hat{\nu}^{(w)} \circ J(dx) \quad (10.97)$$

$$b_0^{(w)} = \hat{b}_0^{(w)} - \frac{1}{2}(\hat{\sigma}^{(w)})^2, \quad (10.98)$$

where

$$J(x) = e^x - 1. \quad (10.99)$$

We next analyze $\sup_w U^{(\alpha)}(r_T(V_\cdot^{(w)})) = -\frac{T}{\alpha} \log\left(\phi_{Z^{(w)}}(i\frac{\alpha}{T}) \right)$. We first note that

$$\sup_w U^{(\alpha)}(r_T(V_\cdot^{(w)})) = \sup_w \left\{ -\frac{T}{\alpha} \log\left(\phi_{Z^{(w)}}(i\frac{\alpha}{T}) \right) \right\}$$

$$= -\frac{T}{\alpha} \log\left(\inf_w \left\{ \phi_{Z^{(w)}}(i\frac{\alpha}{T}) \right\} \right). \quad (10.100)$$

The characteristic function $\phi_{Z^{(w)}}(i\frac{\alpha}{T})$ is computed as follows:

$$
\phi_{Z^{(w)}}\left(i\frac{\alpha}{T}\right)
$$
$$
= \exp\left\{ -\frac{1}{2}(\sigma^{(w)})^2 \left(i\frac{\alpha}{T}\right)^2 + \int_{\Re^1}\left(e^{ix\left(i\frac{\alpha}{T}\right)}-1\right)\nu^{(w)}(dx) + ib_0^{(w)}\left(i\frac{\alpha}{T}\right)\right\}
$$
$$
= \exp\left\{ \frac{1}{2}\left(\frac{\alpha}{T}\right)^2 (\sigma^{(w)})^2 + \int_{\Re^1}\left(e^{-\left(\frac{\alpha}{T}\right)x}-1\right)\nu^{(w)}(dx) - \left(\frac{\alpha}{T}\right)b_0^{(w)}\right\}
$$
$$
= \exp\left\{ \frac{1}{2}\left(\frac{\alpha}{T}\right)^2 \langle \mathbf{w}, \hat{A}\mathbf{w}\rangle + \int_{\Re^1}\left(e^{-\left(\frac{\alpha}{T}\right)x}-1\right)\nu^{(w)}(dx)\right.
$$
$$
\left. - \left(\frac{\alpha}{T}\right)\left(rw^{(0)} + \langle \hat{\mathbf{b}}_0, \mathbf{w}\rangle - \frac{1}{2}\langle \mathbf{w}, \hat{A}\mathbf{w}\rangle\right)\right\}
$$
$$
= \exp\left\{ \int_{\Re^1}\left((1+y)^{-\left(\frac{\alpha}{T}\right)}-1\right)\hat{\nu}^{(w)}(dy)\right.
$$
$$
\left. - \left(\frac{\alpha}{T}\right)\left(rw^{(0)} + \langle \hat{\mathbf{b}}_0, \mathbf{w}\rangle - \frac{1}{2}\left(1+\left(\frac{\alpha}{T}\right)\right)\langle \mathbf{w}, \hat{A}\mathbf{w}\rangle\right)\right\}
$$
$$
= \exp\left\{ \int_{\Re^d}\left((1+\langle \mathbf{w}, \mathbf{x}\rangle)^{-\left(\frac{\alpha}{T}\right)}-1\right)\hat{\nu}(d\mathbf{x})\right.
$$
$$
\left. - \left(\frac{\alpha}{T}\right)\left(rw^{(0)} + \langle \hat{\mathbf{b}}_0, \mathbf{w}\rangle - \frac{1}{2}\left(1+\left(\frac{\alpha}{T}\right)\right)\langle \mathbf{w}, \hat{A}\mathbf{w}\rangle\right)\right\}.
$$
$$
(10.101)
$$

From this we obtain:

$$
\inf_w\left\{\phi_{Z^{(w)}}\left(i\left(\frac{\alpha}{T}\right)\right)\right\}
$$
$$
= \inf_w\left\{\exp\left\{ \int_{\Re^d}\left((1+\langle \mathbf{w}, \mathbf{x}\rangle)^{-\left(\frac{\alpha}{T}\right)}-1\right)\hat{\nu}(d\mathbf{x})\right.\right.
$$
$$
\left.\left. - \left(\frac{\alpha}{T}\right)\left(rw^{(0)} + \langle \hat{\mathbf{b}}_0, \mathbf{w}\rangle - \frac{1}{2}\left(1+\left(\frac{\alpha}{T}\right)\right)\langle \mathbf{w}, \hat{A}\mathbf{w}\rangle\right)\right\}\right\}
$$
$$
= \exp\left\{\inf_w\left\{ \int_{\Re^d}\left((1+\langle \mathbf{w}, \mathbf{x}\rangle)^{-\left(\frac{\alpha}{T}\right)}-1\right)\hat{\nu}(d\mathbf{x})\right.\right.
$$
$$
\left.\left. - \left(\frac{\alpha}{T}\right)\left(rw^{(0)} + \langle \hat{\mathbf{b}}_0, \mathbf{w}\rangle - \frac{1}{2}\left(1+\left(\frac{\alpha}{T}\right)\right)\langle \mathbf{w}, \hat{A}\mathbf{w}\rangle\right)\right\}\right\}.
$$
$$
(10.102)
$$

And from (10.100) and (10.102) we obtain

$$
\begin{aligned}
&\sup_{w} U^{(\alpha)}(r_T(V_{\cdot}^{(w)})) \\
&= -\frac{T}{\alpha} \log \left(\inf_{w} \left\{ \phi_{Z^{(w)}}\left(i\left(\frac{\alpha}{T}\right)\right) \right\} \right) \\
&= -\frac{T}{\alpha} \log \left(\exp \left\{ \inf_{w} \left\{ \int_{\Re^d} \left((1 + \langle \mathbf{w}, \mathbf{x} \rangle)^{-\left(\frac{\alpha}{T}\right)} - 1 \right) \hat{\nu}(d\mathbf{x}) \right. \right. \right. \\
&\qquad\qquad \left. \left. \left. - \left(\frac{\alpha}{T}\right)\left(rw^{(0)} + \langle \hat{\mathbf{b}}_0, \mathbf{w} \rangle - \frac{1}{2}\left(1 + \left(\frac{\alpha}{T}\right)\right) \langle \mathbf{w}, \hat{A}\mathbf{w} \rangle \right) \right\} \right\} \right) \\
&= -\frac{T}{\alpha} \left(\inf_{w} \left\{ \int_{\Re^d} \left((1 + \langle \mathbf{w}, \mathbf{x} \rangle)^{-\left(\frac{\alpha}{T}\right)} - 1 \right) \hat{\nu}(d\mathbf{x}) \right. \right. \\
&\qquad\qquad \left. \left. - \left(\frac{\alpha}{T}\right)\left(rw^{(0)} + \langle \hat{\mathbf{b}}_0, \mathbf{w} \rangle - \frac{1}{2}\left(1 + \left(\frac{\alpha}{T}\right)\right) \langle \mathbf{w}, \hat{A}\mathbf{w} \rangle \right) \right\} \right).
\end{aligned}
$$

$$(10.103)$$

Remark 10.11. From Remark 10.10 and Theorem 10.5, it follows that

$$
\begin{aligned}
\sup_{w} \left\{ U^{(\alpha)}(r_T(V_{\cdot}^{(\theta)})) \right\} &= \sup_{\theta \in \Theta} \left\{ U^{(\alpha)}(r_T(V_{\cdot}^{(\theta)})) \right\} \\
&= -\frac{1}{\alpha}\left(\frac{\alpha + T}{T}\right) \log \left(E\left[\left(\frac{dP^{(ML^q EMM)}}{dP}\right)^q \right] \right).
\end{aligned}
$$
$$(10.104)$$

Remark 10.12. In the case of no jump-part, the optimal strategy is given by

$$
\mathbf{w_T^*} = \frac{\hat{A}^{-1}(\mathbf{b}_0 - r\mathbf{1})}{1 + \left(\frac{\alpha}{T}\right)}.
$$
$$(10.105)$$

10.5.3 *Risk-sensitive evaluation of a single asset*

We consider the risk-sensitive evaluation of a single asset given by

$$
S_t = S_0 e^{Z_t},
$$
$$(10.106)$$

where Z_t is a Lévy process with generating triplet $(\sigma^2, \nu(dx), b_0)_0$. In this case the RIRR of S_t is $r_T(S_{\cdot}) = \frac{1}{T}Z_T$, and the risk-sensitive value of the

RIRR of this asset is

$$U^{(\alpha)}(r_T(S.)) = -\frac{1}{\alpha}\log\left(E\left[e^{-\alpha\left(\frac{1}{T}Z_T\right)}\right]\right) = -\frac{T}{\alpha}\log\left(\phi_Z(i\frac{\alpha}{T})\right)$$

$$= -\frac{T}{\alpha}\log\left(\exp\left\{-\frac{1}{2}\sigma^2\left(i\frac{\alpha}{T}\right)^2 + \int_{\Re^1}\left(e^{ix\left(i\frac{\alpha}{T}\right)} - 1\right)\nu(dx) + ib_0\left(i\frac{\alpha}{T}\right)\right\}\right)$$

$$= \frac{T}{\alpha}\left(-\frac{1}{2}\sigma^2\left(\frac{\alpha}{T}\right)^2 - \int_{\Re^1}\left(e^{-x\left(\frac{\alpha}{T}\right)} - 1\right)\nu(dx) + b_0\left(\frac{\alpha}{T}\right)\right)$$

$$= b_0 - \frac{\alpha}{2T}\sigma^2 - \frac{T}{\alpha}\int_{\Re^1}\left(e^{-x\left(\frac{\alpha}{T}\right)} - 1\right)\nu(dx). \tag{10.107}$$

Notes

For the details of d-dimensional Lévy processes, see Sato (1999) [109]. For the relation between the Esscher martingale measure (ESSMM) and the minimal entropy martingale measure (MEMM), see Hubalek and Sgarra (2006) [53].

For portfolio problems there are many papers and books. For the evaluation problem of portfolios, see Miyahara and Tsujii (2011) [96]. The results of Sections 10.4 and 10.5 are based on this paper.

Appendix A

Estimation

Lévy process models are one of the most important classes of stochastic models in mathematical finance. To apply a Lévy process model to the empirical analysis of financial problems, we have to solve the estimation problem of Lévy processes.

We will adopt the generalized method of moments for the estimation of Lévy processes. In this chapter we will explain this method briefly.

A.1 Method of Moments

We will first survey the method of moments.

A.1.1 *Classical method of moments*

1) Moments of random variables:

Let X be a random variable defined on some probability space. Then the k-th moment of X in the sense of distribution, which is denoted by m_k, is defined as

$$m_k = E[X^k], \quad k = 1, 2, \ldots \tag{A.1}$$

This is also called the *k-th moment of X*.

2) Sample moment:

A sample of X with n-length is a sequence of independent, identically distributed (i.i.d.) random variables $\{X_j, j = 1, \ldots, n\}$, whose distribution is equal to that of X. In the above situation, the *k-th sample moment* is

$$\overline{(X^k)_n} = \frac{1}{n} \sum_{j=1}^{n} (X_j)^k, \quad k = 1, 2, \ldots \tag{A.2}$$

It holds that

$$E\left[\overline{(X^k)_n}\right] = E[X^k] = m_k. \tag{A.3}$$

3) Moments of sample data:

Let $\{\xi_1, \ldots, \xi_n\}$ be a sequence of sample data of X. Then the k-th moment of sample data, $\hat{m}_k$, is defined as

$$\hat{m}_k = \frac{1}{n} \sum_{j=1}^{n} (\xi_j)^k, \quad k = 1, 2, \ldots \tag{A.4}$$

4) Classical moment-equation:

Assume that the distribution of X contains δ parameters $\{\beta_1, \ldots, \beta_\delta\}$, and assume that a sample data of X $\{\xi_1, \ldots, \xi_n\}$ are given. The problem to be solved is to estimate the true values of parameters from the data.

Assume that the moments $m_k = E[X^k], k = 1, 2, \ldots$, are finite. Then, applying the law of large numbers to a sample of X, $\{X_j, j = 1, \ldots\}$, we can obtain the following formula with probability one:

$$\lim_{n \to \infty} \overline{(X^k)_n} = \lim_{n \to \infty} \frac{1}{n} \sum_{j=1}^{n} (X_j)^k = m_k, \quad k = 1, 2, \ldots \tag{A.5}$$

From this we know that almost all sample paths $\{\xi_1, \xi_2, \ldots\}$ satisfy the following equality:

$$\lim_{n \to \infty} \hat{m}_k = \lim_{n \to \infty} \frac{1}{n} \sum_{j=1}^{n} (\xi_j)^k = m_k, \quad k = 1, 2, \ldots \tag{A.6}$$

We remark here that m_k is dependent on the parameters $\{\beta_1, \ldots, \beta_\delta\}$.

The classical method of moments is based on the above fact. Let the sample number n be large enough, and consider the following equation with respect to the parameters $\{\beta_1, \ldots, \beta_\delta\}$:

$$m_k(\beta_1, \ldots, \beta_\delta) = \hat{m}_k, \quad k = 1, 2, \ldots, \delta. \tag{A.7}$$

This equation is called the "moment-equation", and the solution of this equation is an estimator of the parameters $\{\beta_1, \ldots, \beta_\delta\}$. This estimation method is called the "classical method of moments".

5) Characteristic function:

The characteristic function $\phi_X(u)$ of X is defined by

$$\phi_X(u) = e^{\psi_X(u)} = E[e^{iuX}], \quad i = \sqrt{-1}. \tag{A.8}$$

We use the following notations:

$$\phi^{(k)}(u) = \frac{d^k \phi_X}{du^k}(u), \quad \psi^{(k)}(u) = \frac{d^k \psi}{du^k}(u). \tag{A.9}$$

It is well known that, if X has k-th moments, then the following equality holds:

$$m_k = E[X^k] = \frac{1}{i^k} E[(iX)^k] = \frac{1}{i^k} \phi_X^{(k)}(0), \quad k = 1, 2, \ldots \tag{A.10}$$

So the classical moment-equations are

$$\frac{1}{i^k} \phi_X^{(k)}(0) = \hat{m}_k, \quad k = 1, \ldots, \delta, \tag{A.11}$$

where $\phi_X^{(k)}(0)$ is a function of the parameters $\{\beta_1, \ldots, \beta_\delta\}$.

6) Transformation of the moment-equation:

We will now consider only the case of $\delta = 4$. Using the following formulas:

$$\phi_X^{(1)}(u) = \psi^{(1)}(u)\phi_X(u), \tag{A.12}$$

$$\phi_X^{(2)}(u) = \left(\psi_X^{(2)}(u) + (\psi_X^{(1)}(u))^2 \right) \phi_X(u), \tag{A.13}$$

$$\phi_X^{(3)}(u) = \left(\psi_X^{(3)}(u) + 3\psi_X^{(2)}(u)\psi_X^{(1)}(u) + (\psi_X^{(1)}(u))^3 \right) \phi_X(u), \tag{A.14}$$

$$\phi_X^{(4)}(u) = \left(\psi_X^{(4)}(u) + 4\psi_X^{(3)}(u)\psi_X^{(1)}(u) + 3(\psi_X^{(2)}(u))^2 \right.$$
$$\left. + 6\psi_X^{(2)}(u)(\psi_X^{(1)}(u))^2 + (\psi_X^{(1)}(u))^4 \right) \phi_X(u), \tag{A.15}$$

we obtain the following results:

$$m_1 = \frac{1}{i} \psi_X^{(1)}(0), \tag{A.16}$$

$$m_2 = - \left(\psi_X^{(2)}(0) + (\psi_X^{(1)}(0))^2 \right), \tag{A.17}$$

$$m_3 = -\frac{1}{i} \left(\psi_X^{(3)}(0) + 3\psi_X^{(2)}(0)\psi_X^{(1)}(0) + (\psi_X^{(1)}(0))^3 \right), \tag{A.18}$$

$$m_4 = \psi_X^{(4)}(0) + 4\psi_X^{(3)}(0)\psi_X^{(1)}(0) + 3(\psi_X^{(2)}(0))^2$$
$$+ 6\psi_X^{(2)}(0)(\psi_X^{(1)}(0))^2 + (\psi_X^{(1)}(0))^4. \tag{A.19}$$

From these results we obtain:

$$\psi_X^{(1)}(0) = im_1, \tag{A.20}$$

$$\begin{aligned}\psi_X^{(2)}(0) &= -m_2 - (\psi_X^{(1)}(0))^2 \\ &= -m_2 - (im_1)^2 \\ &= -(m_2 - m_1^2),\end{aligned} \tag{A.21}$$

$$\begin{aligned}\psi_X^{(3)}(0) &= -im_3 - \left(3\psi_X^{(2)}(0)\psi_X^{(1)}(0) + (\psi_X^{(1)}(0))^3\right) \\ &= -im_3 - 3(-m_2 + m_1^2)im_1 - (im_1)^3 \\ &= -i(m_3 - 3m_2m_1 + 2m_1^3),\end{aligned} \tag{A.22}$$

$$\begin{aligned}\psi_X^{(4)}(0) &= m_4 - \left(4\psi_X^{(3)}(0)\psi_X^{(1)}(0) + 3(\psi_X^{(2)}(0))^2\right. \\ &\qquad \left. +6\psi_X^{(2)}(0)(\psi_X^{(1)}(0))^2 + (\psi_X^{(1)}(0))^4\right) \\ &= m_4 - 4(i(-m_3 + 3m_2m_1 - 2m_1^3)im_1) - 3(-m_2 + m_1^2)^2 \\ &\quad -6(-m_2 + m_1^2)(im_1)^2 - (im_1)^4 \\ &= m_4 - 4m_3m_1 - 3m_2^2 + 12m_2m_1^2 - 6m_1^4.\end{aligned} \tag{A.23}$$

On the other hand, let us suppose that a sequence of sample data of X, $\{\xi_1, \ldots, \xi_n\}$, is given. Let $\hat{m}_k$ be the k-th moment of sample data defined by (A.4), and set:

$$\hat{h}_1 = \hat{m}_1, \tag{A.24}$$
$$\hat{h}_2 = \hat{m}_2 - \hat{m}_1^2, \tag{A.25}$$
$$\hat{h}_3 = \hat{m}_3 - 3\hat{m}_2\hat{m}_1 + 2\hat{m}_1^3, \tag{A.26}$$
$$\hat{h}_4 = \hat{m}_4 - 4\hat{m}_3\hat{m}_1 - 3\hat{m}_2^2 + 12\hat{m}_2\hat{m}_1^2 - 6\hat{m}_1^4. \tag{A.27}$$

Then the moment-equation (A.11) is identified with the following equations:

$$\psi_X^{(1)}(0) = i\hat{h}_1, \tag{A.28}$$
$$\psi_X^{(2)}(0) = -\hat{h}_2, \tag{A.29}$$
$$\psi_X^{(3)}(0) = -i\hat{h}_3, \tag{A.30}$$
$$\psi_X^{(4)}(0) = \hat{h}_4. \tag{A.31}$$

This is the third form of classical moment-equation.

Definition A.1. The equations obtained above are called "transformed moment-equations".

A.1.2 Generalized method of moments (= characteristic function method)

The idea of the classical method of moments, which is described above, is developed into the generalized method of moments. Let $\phi_X(u)$ be the characteristic function of X. If we could estimate $\phi_X(u)$, then the distribution of X would be estimated. Let $\{\beta_1, \ldots, \beta_\delta\}$ be parameters of distribution of X. The problem to solve is the estimation of these parameters.

1) Sample characteristic function:

Let $\{X_j, j = 1, \ldots, n\}$ be a sample of X with n-length, and set

$$Y_j^{(u)} = e^{iuX_j}, \quad -\infty < u < \infty, \quad j = 1, 2, \ldots \tag{A.32}$$

Then $Y_j, j = 1, 2, \ldots$, is a sequence of i.i.d. random variables, whose distribution is equal to that of e^{iuX}. Applying the law of large numbers to $Y_j^{(u)}, j = 1, 2, \ldots$, we can obtain

$$\lim_{n\to\infty} \overline{(Y^{(u)})_n} = \lim_{n\to\infty} \frac{1}{n} \sum_{j=1}^{n} e^{iuX_j} = E[e^{iuX}] = \phi_X(u), \quad -\infty < u < \infty,$$
$$\tag{A.33}$$

with probability one. We call $\frac{1}{n} \sum_{j=1}^{n} e^{iuX_j}$ the "sample characteristic function" of X. It holds that

$$E\left[\overline{(Y^{(u)})_n}\right] = E[e^{iuX_j}] = E[e^{iuX}] = \phi_X(u), \quad -\infty < u < \infty. \tag{A.34}$$

2) Characteristic function of sample data:

Let $\{\xi_1, \ldots, \xi_n\}$ be a sequence of sample data of X. Then we define the characteristic function of sample data, $\hat{\phi}_X(u)$, by

$$\hat{\phi}_X(u) = \frac{1}{n} \sum_{j=1}^{n} e^{iu\xi_j}. \tag{A.35}$$

From (A.33) we know that $\hat{\phi}_X(u)$ is a consistent estimator of $\phi_X(u)$:

$$\lim_{n\to\infty} \hat{\phi}_X(u) = \phi_X(u), \quad -\infty < u < \infty. \tag{A.36}$$

3) Generalized moment-equation:

The following equation is called the "generalized moment-equation":

$$\phi_X(u) = \hat{\phi}_X(u) = \frac{1}{n} \sum_{j=1}^{n} e^{iu\xi_j}, \quad -\infty < u < \infty. \tag{A.37}$$

It should be noted that $\phi_X(u)$ is depending on parameter $\{\beta_1, \ldots, \beta_\delta\}$.

A.1.3 *Estimation of Lévy processes*

The Lévy process Z_t is characterized by the generating triplet $(\sigma^2, \nu(dx), b)$ (or $(\sigma^2, \nu(dx), b_c)_c)$. Set $Z = Z_1$. From Theorem 2.2 the distribution of Z is an infinitely divisible distribution, and from Theorem 2.3 the corresponding characteristic function $\phi_Z(u)$ is

$$
\phi_Z(u) = E[e^{iuZ}] = \exp(\psi_Z(u))
$$

$$
= \exp\left\{ -\frac{\sigma^2}{2}u^2 + \int_{|x|<1} (e^{iux} - 1 - iux)\nu(dx) \right.
$$

$$
\left. + \int_{|x|\geq 1} (e^{iux} - 1)\nu(dx) + ibu \right\} \tag{A.38}
$$

$$
= \exp\left\{ -\frac{\sigma^2}{2}u^2 + \int_{(-\infty,\infty)} \left(e^{iux} - 1 - iuxc(x)\right)\nu(dx) + ib_cu \right\}.
$$

$$
\tag{A.39}
$$

Since the distribution of an infinitely divisible distribution is uniquely determined by its generating triplet, we need to estimate the generating triplet $(\sigma^2, \nu(dx), b)$ (or $(\sigma^2, \nu(dx), b_c)_c)$ of the distribution of Z. Set $\{\tilde{Z}_j = Z_j - Z_{j-1}, j = 1, 2, \ldots\}$, then $\{\tilde{Z}_j, j = 1, 2, \ldots\}$ is i.i.d. with the same distribution as Z, since Lévy processes have temporally homogeneous independent increments. So, if we are given the sequential data of a Lévy process, then we can apply the classical method of moments or generalized method of moments (= characteristic function method), which are described above, to the estimation problem of Z.

Let $\{z_j, j = 1, 2, \ldots, n\}$ be a sequential data set corresponding to $\{Z_j, j = 1, 2, \ldots, n\}$. Then $\{\xi_j = z_j - z_{j-1}, j = 1, 2, \ldots, n\}$ is a sequential data set corresponding to $\{\tilde{Z}_j = Z_j - Z_{j-1}, j = 1, 2, \ldots, n\}$, namely $\{\xi_j, j = 1, 2, \ldots, n\}$ is a sequence of sample data set of Z. Using this sequence of sample data $\{\xi_j, j = 1, 2, \ldots, n\}$, we obtain the classical moment-equations (A.7) and (A.11) for the distribution of Z:

$$
m_k(\beta_1, \ldots, \beta_\delta) = \hat{m}_k, \quad k = 1, 2, \ldots, \delta, \tag{A.40}
$$

and

$$
\frac{1}{i^k}\phi_Z^{(k)}(0) = \hat{m}_k, \quad k = 1, \ldots, \delta. \tag{A.41}
$$

The transformed moment-equations described in 6) of Section B.1.1 are:

$$\psi_Z^{(1)}(0) = i\hat{h}_1, \tag{A.42}$$

$$\psi_Z^{(2)}(0) = -\hat{h}_2, \tag{A.43}$$

$$\psi_Z^{(3)}(0) = -i\hat{h}_3, \tag{A.44}$$

$$\psi_Z^{(4)}(0) = \hat{h}_4. \tag{A.45}$$

The generalized moment-equation described in 3) of Section B.1.2. is:

$$\phi_Z(u) = \hat{\phi}_Z(u) = \frac{1}{n}\sum_{j=1}^{n} e^{iu\xi_j}, \quad -\infty < u < \infty. \tag{A.46}$$

A.2 Examples

In this section we see some examples of the estimation of Lévy processes. Our method is the classical or generalized method of moments described in the previous section.

A.2.1 *Compound Poisson process with normal Lévy measures*

For simplicity and utility, we will consider only the case of normal Lévy measures. (Other cases can be treated in the same manner.) So we suppose that the Lévy measure is

$$\nu(dx) = cg(x; m, v)dx = c\frac{1}{\sqrt{2\pi v}}\exp\left(-\frac{(x-m)^2}{2v}\right)dx, \tag{A.47}$$

and the characteristic function $\phi(u)$ is of the following form with the parameters (c, m, v):

$$\phi(u) = E[e^{iuZ}] = \exp(\psi(u)), \tag{A.48}$$

$$\psi(u) = \int_{-\infty}^{\infty} (e^{iux} - 1)\nu(dx) = c\left(\hat{g}(u) - 1\right), \tag{A.49}$$

where

$$\hat{g}(u) = \int_{-\infty}^{\infty} e^{iux} g(x; m, v)dx = \exp\left(imu - \frac{1}{2}vu^2\right). \tag{A.50}$$

We can apply the classical method of moments described in Section B.1 to estimate the parameters (c, m, v).

It is easy to see that

$$\psi^{(1)}(u) = c(im - vu)\hat{g}(u), \tag{A.51}$$

$$\psi^{(2)}(u) = \left(-cv + c(im - vu)^2\right)\hat{g}(u), \tag{A.52}$$

$$\psi^{(3)}(u) = \left(-3cv(im - vu) + c(im - vu)^3\right)\hat{g}(u). \tag{A.53}$$

From this it follows that

$$\psi^{(1)}(0) = cm, \tag{A.54}$$

$$\psi^{(2)}(0) = -c(v + m^2), \tag{A.55}$$

$$\psi^{(3)}(0) = -ic(3v + m^2)m. \tag{A.56}$$

Therefore the transformed moment-equations are

$$cm = \hat{h}_1 \tag{A.57}$$

$$c(v + m^2) = \hat{h}_2 \tag{A.58}$$

$$c(3v + m^2)m = \hat{h}_3. \tag{A.59}$$

(See Definition A.1.)

Solving these equations under the condition that $c > 0$ and $v \geq 0$, we obtain an estimator $(\hat{c}, \hat{m}, \hat{v})$.

Remark A.1. $\hat{m}$ is a solution of

$$2m^2 - 3\frac{\hat{h}_2}{\hat{h}_1}m + \frac{\hat{h}_3}{\hat{h}_1} = 0, \tag{A.60}$$

and $\hat{c}$ and $\hat{v}$ are given as follows:

$$\hat{c} = \frac{\hat{h}_1}{\hat{m}}, \tag{A.61}$$

$$\hat{v} = -\hat{m}^2 + \frac{\hat{h}_2}{\hat{h}_1}\hat{m}. \tag{A.62}$$

A.2.2 *Jump-diffusion process (compound Poisson diffusion process)*

Suppose that the Lévy process Z_t is of the following form:

$$Z_t = \sigma W_t + b_0 t + J_t, \tag{A.63}$$

where J_t is a compound Poisson process. Then the generating triplet of Z_t is $(\sigma^2, \nu(dx), b_0)_0$, and the Lévy measure $\nu(dx)$ is

$$\nu(dx) = c\rho(dx), \tag{A.64}$$

where c is a positive constant and $\rho(dx)$ is a probability on $(-\infty, \infty)$ such that $\rho(\{0\}) = 0$. Then the characteristic function $\phi(u)$ is of the following form:

$$\phi(u) = E[e^{iuZ}] = \exp(\psi(u)), \tag{A.65}$$

$$\psi(u) = -\frac{1}{2}\sigma^2 u^2 + ib_0 u + c\left(\hat{\rho}(u) - 1\right), \tag{A.66}$$

where

$$\hat{\rho}(u) = \int_{-\infty}^{\infty} e^{iux}\rho(dx). \tag{A.67}$$

We mention here that $|\hat{\rho}(u)| \leq 1$, so

$$\left| c\left(\hat{\rho}(u) - 1\right) \right| \leq 2c. \tag{A.68}$$

From this we obtain the following propositions.

Proposition A.1.

$$-2 \lim_{|u| \to \infty} \frac{Re[\psi(u)]}{u^2} = \sigma^2. \tag{A.69}$$

Proposition A.2.

$$\lim_{|u| \to \infty} \frac{Im[\psi(u)]}{u} = b_0. \tag{A.70}$$

These propositions are effective for the estimation of σ^2 and b_0. Suppose that a sequential data set of Z_t, $\{z_1, \ldots, z_n\}$, is given and let $\hat{\phi}(u)$ be the characteristic function of sample data

$$\hat{\phi}(u) = \frac{1}{n}\sum_{j=1}^{n} e^{iu\xi_j}, \quad \xi_j = z_j - z_{j-1}, \tag{A.71}$$

and set

$$\hat{\psi}(u) = \log(\hat{\phi}(u)). \tag{A.72}$$

From (A.36) and the above propositions, we can calculate that when u is large enough, then

$$-2\frac{Re[\hat{\psi}(u)]}{u^2} \sim \sigma^2, \tag{A.73}$$

and

$$\frac{Im[\hat{\psi}(u)]}{u} \sim b_0. \tag{A.74}$$

Thus we have obtained the following estimators.

Select a large number u_1 and set

$$\hat{\sigma}^2 = -2\frac{Re[\hat{\psi}(u_1)]}{u_1^2}, \tag{A.75}$$

and

$$\hat{b}_0 = \frac{Im[\hat{\psi}(u_1)]}{u_1}. \tag{A.76}$$

Then $\hat{\sigma}^2$ is an estimator of σ^2 and $\hat{b}_0$ is an estimator of b_0.

Example A.1. (Discrete Lévy measure). The parameters σ^2 and b_0 are estimated by (A.75) and (A.76) respectively. Suppose that the Lévy measure $\nu(dx)$ is of the following form:

$$\nu(dx) = c\rho(dx) = c\sum_{j=1}^{d} p_j \delta_{a_j}(dx), \quad p_j \geq 0, j = 1, 2, \ldots, d, \quad \sum_{j=1}^{d} p_j = 1. \tag{A.77}$$

Then

$$\psi(u) = -\frac{1}{2}\sigma^2 u^2 + ib_0 u + c\sum_{j=1}^{d} p_j(e^{iua_j} - 1). \tag{A.78}$$

The parameters c and $p_j, j = 1, \ldots, d$ are estimated by the $d+1$ transformed moment-equations. (See Definition A.1.)

Example A.2. (Normal Lévy measure). The estimation of σ^2 and b_0 is the same as in the above example. When we have obtained estimators $\hat{\sigma}^2$ and $\hat{b}_0$, the rest is continued in the same way as in Section B.2.1.

A.2.3   *Stable process*

We start from the characteristic function of a stable process:

$$\phi(u) = \exp(\psi(u)),$$

$$\psi(u) = \begin{cases} -c|u|^\alpha \left(1 - i\beta\tan\frac{\pi\alpha}{2}\mathrm{sgn}(u)\right) + i\tau u, & for \quad \alpha \neq 1 \\ -c|u| \left(1 + i\beta\frac{2}{\pi}\mathrm{sgn}(u)\log|u|\right) + i\tau u, & for \quad \alpha = 1. \end{cases} \tag{A.79}$$

where (α, c, β, τ) are the parameters of a stable process such that

$$0 < \alpha < 2, \quad c > 0, \quad -1 \leq \beta \leq 1, \quad -\infty < \tau < \infty. \tag{A.80}$$

Remark A.2. Comparing this expression of $\psi(u)$ with the expression (2.42), we know that

$$c = \begin{cases} -\Gamma(-\alpha)(\cos\frac{\pi\alpha}{2})\tilde{c}, & for \quad \alpha \neq 1 \\ \frac{\pi}{2}\tilde{c}, & for \quad \alpha = 1. \end{cases} \tag{A.81}$$

We introduce the polar coordinate of $\phi(u)$:

$$\phi(u) = \rho(u)e^{i\theta(u)}, \tag{A.82}$$

where

$$\rho(u) = |\phi(u)| = e^{-c|u|^\alpha}, \tag{A.83}$$

and

$$\theta(u) = \text{Im}[\psi(u)] = \begin{cases} c|u|^\alpha \beta \tan \frac{\pi\alpha}{2} \text{sgn}(u) + \tau u, & for \quad \alpha \neq 1 \\ -c|u|\beta \frac{2}{\pi} \text{sgn}(u) \log|u| + \tau u, & for \quad \alpha = 1. \end{cases} \tag{A.84}$$

We first check whether $\alpha = 1$ or $\alpha \neq 1$.

Judgment of $\alpha = 1$ or $\alpha \neq 1$:
From (A.83) it follows that

$$\log|\phi(u)| = -c|u|^\alpha. \tag{A.85}$$

Therefore if $\alpha = 1$ then $\log|\phi(u)|$ is a linear function of u on $(-\infty, 0)$ or on $(0, \infty)$. Since $\hat{\phi}(u) \sim \phi(u)$, it follows that $\log|\hat{\phi}(u)|$ is an almost linear function of u on $(-\infty, 0)$ or on $(0, \infty)$. Using this fact, we can judge whether $\alpha = 1$ or not.

(i) Case of $\alpha \neq 1$.
Assume that we have judged that $\alpha \neq 1$. Choose u_1 and u_2 such that $u_1, u_2 \neq 0$ and $u_1 \neq u_2$, and we use the generalized moment-equation (A.37) which is equivalent to the following equation:

$$|\phi(u)| = |\hat{\phi}(u)|, \tag{A.86}$$

for $u = u_1$ and $u = u_2$. Then we obtain the following equations:

$$c|u_1|^\alpha = -\log|\hat{\phi}(u_1)|, \qquad c|u_2|^\alpha = -\log|\hat{\phi}(u_2)|, \tag{A.87}$$

where we remark that $-\log|\hat{\phi}(u)| \geq 0$ because of $|\hat{\phi}(u)| \leq 1$. Solving this equation, we obtain the following estimators:

$$\hat{\alpha} = \frac{\log\left(\frac{\log|\hat{\phi}(u_1)|}{\log|\hat{\phi}(u_2)|}\right)}{\log\left|\frac{u_1}{u_2}\right|}, \tag{A.88}$$

and

$$\hat{c} = -\frac{\log|\hat{\phi}(u_1)|}{|u_1|^{\hat{\alpha}}}. \tag{A.89}$$

For the estimation of β and τ, we use the polar expression (A.83). Set the following:

$$\hat{\theta}(u) = \text{Im}\hat{\psi}(u) = \text{Im}(\text{Log}\hat{\phi}(u)), \tag{A.90}$$

and choose u_3 and u_4 such that $u_3, u_4 \neq 0$ and $u_3 \neq u_4$. Using the moment-equation $\theta(u) = \hat{\theta}(u)$, which follows from the generalized moment-equation (A.37), we obtain the following equations:

$$\hat{c}|u_3|^{\hat{\alpha}}\beta \tan \frac{\pi\hat{\alpha}}{2}\mathrm{sgn}(u_3) + \tau u_3 = \hat{\theta}(u_3), \tag{A.91}$$

$$\hat{c}|u_4|^{\hat{\alpha}}\beta \tan \frac{\pi\hat{\alpha}}{2}\mathrm{sgn}(u_4) + \tau u_4 = \hat{\theta}(u_4). \tag{A.92}$$

These are linear equations for β and τ, and the solution is

$$
\begin{aligned}
\hat{\beta} &= \frac{\hat{\theta}(u_3)u_4 - \hat{\theta}(u_4)u_3}{\hat{c}(\tan \frac{\pi\hat{\alpha}}{2})\left(|u_3|^{\hat{\alpha}}\mathrm{sgn}(u_3)u_4 - |u_4|^{\hat{\alpha}}\mathrm{sgn}(u_4)u_3\right)} \\
&= \frac{\hat{\theta}(u_3)u_4 - \hat{\theta}(u_4)u_3}{\hat{c}(\tan \frac{\pi\hat{\alpha}}{2})(|u_3|^{\hat{\alpha}-1} - |u_4|^{\hat{\alpha}-1})u_3 u_4},
\end{aligned}
\tag{A.93}
$$

and

$$
\begin{aligned}
\hat{\tau} &= \frac{|u_3|^{\hat{\alpha}}\mathrm{sgn}(u_3)\hat{\theta}(u_4) - |u_4|^{\hat{\alpha}}\mathrm{sgn}(u_4)\hat{\theta}(u_4)}{|u_3|^{\hat{\alpha}}\mathrm{sgn}(u_3)u_4 - |u_4|^{\hat{\alpha}}\mathrm{sgn}(u_4)u_3} \\
&= \frac{|u_3|^{\hat{\alpha}-1}u_3\hat{\theta}(u_4) - |u_4|^{\hat{\alpha}-1}u_4\hat{\theta}(u_4)}{(|u_3|^{\hat{\alpha}-1} - |u_4|^{\hat{\alpha}-1})u_3 u_4}.
\end{aligned}
\tag{A.94}
$$

(ii) Case of $\alpha = 1$.

Assume that we have judged that $\alpha = 1$, and choose $u_1 \neq 0$. From (A.83) and the generalized moment-equation, we obtain an estimator of c:

$$\hat{c} = -\frac{\log |\phi(u_1)|}{|u_1|}. \tag{A.95}$$

Using (A.84), (A.95), and $\alpha = 1$, from the moment-equation $\theta(u) = \hat{\theta}(u)$ we obtain the following equations:

$$\left(-\hat{c}\beta\frac{2}{\pi}\log |u_3| + \tau\right) u_3 = \hat{\theta}(u_3), \tag{A.96}$$

and

$$\left(-\hat{c}\beta\frac{2}{\pi}\log |u_4| + \tau\right) u_4 = \hat{\theta}(u_4). \tag{A.97}$$

Solving these equations, we obtain the following estimators:

$$\hat{\beta} = \frac{\hat{\theta}(u_3)u_4 - \hat{\theta}(u_4)u_3}{-\hat{c}\frac{2}{\pi}(\log |u_3| - \log |u_4|)u_3 u_4}, \tag{A.98}$$

$$\hat{\tau} = \frac{(\log |u_3|)u_3\hat{\theta}(u_4) - (\log |u_4|)u_4\hat{\theta}(u_3)}{(\log |u_3| - \log |u_4|)u_3 u_4}. \tag{A.99}$$

A.2.4 *Variance gamma process*

A variance gamma process is determined by four parameters (C, c_1, c_2, b_0), where C, c_1, c_2 are positive and $-\infty < b_0 < \infty$. The Lévy measure is given by

$$\nu(dx) = C \left(I_{\{x<0\}} \exp(-c_1|x|) + I_{\{x>0\}} \exp(-c_2|x|) \right) |x|^{-1} dx, \quad \text{(A.100)}$$

and the generating triplet is $(0, \nu(dx), b_0)_0$. The characteristic function is

$$\phi_{VG}(u) = \exp\{\psi(u)\}$$
$$= \exp\left\{ ib_0 u - C \left(\log(1 + \frac{iu}{c_1}) + \log(1 - \frac{iu}{c_2}) \right) \right\} \quad \text{(A.101)}$$
$$= e^{ib_0 u} \left(\frac{1}{\left(1 + \frac{iu}{c_1}\right)\left(1 - \frac{iu}{c_2}\right)} \right)^C. \quad \text{(A.102)}$$

A.2.4.1 *Estimation by classical method of moments*

From (A.101) it follows that:

$$\psi^{(1)}(u) = ib_0 + iC \left(\frac{1}{c_2 - iu} - \frac{1}{c_1 - iu} \right), \quad \text{(A.103)}$$

$$\psi^{(2)}(u) = -C \left(\frac{1}{(c_2 - iu)^2} + \frac{1}{(c_1 - iu)^2} \right), \quad \text{(A.104)}$$

$$\psi^{(3)}(u) = -2iC \left(\frac{1}{(c_2 - iu)^3} - \frac{1}{(c_1 - iu)^3} \right), \quad \text{(A.105)}$$

$$\psi^{(4)}(u) = 6C \left(\frac{1}{(c_2 - iu)^4} + \frac{1}{(c_1 - iu)^4} \right). \quad \text{(A.106)}$$

So we obtain:

$$\psi^{(1)}(0) = i \left(b_0 + C \left(\frac{1}{c_2} - \frac{1}{c_1} \right) \right), \quad \text{(A.107)}$$

$$\psi^{(2)}(0) = -C \left(\frac{1}{c_2^2} + \frac{1}{c_1^2} \right), \quad \text{(A.108)}$$

$$\psi^{(3)}(0) = -i2C \left(\frac{1}{c_2^3} - \frac{1}{c_1^3} \right), \quad \text{(A.109)}$$

$$\psi^{(4)}(0) = 6C \left(\frac{1}{c_2^4} + \frac{1}{c_1^4} \right). \quad \text{(A.110)}$$

From the above results, the transformed moment-equations are:

$$b_0 + C\left(\frac{1}{c_2} - \frac{1}{c_1}\right) = \hat{h}_1, \tag{A.111}$$

$$C\left(\frac{1}{c_2^2} + \frac{1}{c_1^2}\right) = \hat{h}_2, \tag{A.112}$$

$$2C\left(\frac{1}{c_2^3} - \frac{1}{c_1^3}\right) = \hat{h}_3, \tag{A.113}$$

$$6C\left(\frac{1}{c_2^4} + \frac{1}{c_1^4}\right) = \hat{h}_4. \tag{A.114}$$

We can solve these equations as follows. Set the following:

$$x_1 = -\frac{1}{c_1} < 0, \quad x_2 = \frac{1}{c_2} > 0, \tag{A.115}$$

then

$$x_1 + x_2 = \frac{\hat{h}_1 - b_0}{C}, \tag{A.116}$$

$$x_1^2 + x_2^2 = \frac{\hat{h}_2}{C}, \tag{A.117}$$

$$x_1^3 + x_2^3 = \frac{\hat{h}_3}{C}, \tag{A.118}$$

$$x_1^4 + x_2^4 = \frac{\hat{h}_4}{C}. \tag{A.119}$$

From (A.117) we obtain

$$(x_1 + x_2)^2 = x_1^2 + x_2^2 + 2x_1x_2 = \frac{\hat{h}_2}{C} + 2x_1x_2, \tag{A.120}$$

and from (A.116) and (A.120) we get

$$2x_1x_2 = \frac{1}{2}\left(\frac{(\hat{h}_1 - b_0)^2}{C^2} - \frac{\hat{h}_2}{C}\right). \tag{A.121}$$

When b_0 and C are given, by (A.116) and (A.121), x_1 and x_2 are solutions of the following equation:

$$x^2 - \frac{(\hat{h}_1 - b_0)}{C}x + \frac{1}{2}\left(\frac{(\hat{h}_1 - b_0)^2}{C^2} - \frac{\hat{h}_2}{C}\right) = 0. \tag{A.122}$$

Next we derive the equations for C and b_0. Using (A.116) and (A.117), we obtain

$$(x_1 + x_2)^3 = (x_1 + x_2)(x_1^2 - x_1x_2 + x_2^2)$$

$$= \frac{(\hat{h}_1 - b_0)}{C}\left(\frac{\hat{h}_2}{C} - \frac{1}{2}\left(\frac{(\hat{h}_1 - b_0)^2}{C^2} - \frac{\hat{h}_2}{C}\right)\right)$$

$$= \frac{(\hat{h}_1 - b_0)}{C}\left(\frac{3}{2}\frac{\hat{h}_2}{C} - \frac{1}{2}\frac{(\hat{h}_1 - b_0)^2}{C^2}\right). \tag{A.123}$$

So, from (A.118) it follows that

$$\hat{h}_3 = (\hat{h}_1 - b_0) \left(\frac{3}{2} \frac{\hat{h}_2}{C} - \frac{1}{2} \frac{(\hat{h}_1 - b_0)^2}{C^2} \right), \qquad \text{(A.124)}$$

and

$$\hat{h}_3 C^2 - 3\hat{h}_2(\hat{h}_1 - b_0)C + (\hat{h}_1 - b_0)^3 = 0. \qquad \text{(A.125)}$$

Using (A.117) and (A.121), we obtain

$$\begin{aligned}
x_1^4 + x_2^4 &= (x_1^2 + x_2^2)^2 - 2x_1^2 x_2^2 \\
&= \left(\frac{\hat{h}_2}{C} \right)^2 - 2 \left(\frac{1}{2} \left(\frac{(\hat{h}_1 - b_0)^2}{C^2} - \frac{\hat{h}_2}{C} \right) \right)^2 \\
&= \frac{1}{2} \frac{\hat{h}_2^2}{C^2} + \frac{\hat{h}_2(\hat{h}_1 - b_0)^2}{C^3} - \frac{1}{2} \frac{(\hat{h}_1 - b_0)^4}{C^4}.
\end{aligned} \qquad \text{(A.126)}$$

It follows from (A.119) that

$$\hat{h}_4 = \frac{1}{2} \frac{\hat{h}_2^2}{C} + \frac{\hat{h}_2(\hat{h}_1 - b_0)^2}{C^2} - \frac{1}{2} \frac{(\hat{h}_1 - b_0)^4}{C^3}, \qquad \text{(A.127)}$$

and

$$\hat{h}_4 C^3 - \frac{1}{2}\hat{h}_2^2 C^2 - \hat{h}_2(\hat{h}_1 - b_0)^2 C + \frac{1}{2}(\hat{h}_1 - b_0)^4 = 0. \qquad \text{(A.128)}$$

Computing $(A.128) - (A.125) \times (\hat{h}_1 - b_0)$, we obtain:

$$\hat{h}_2(\hat{h}_1 - b_0)^2 - 2\hat{h}_3 C(\hat{h}_1 - b_0) - \hat{h}_2^2 C + 2\hat{h}_4 C^2 = 0. \qquad \text{(A.129)}$$

Eliminating $(\hat{h}_1 - b_0)$ by computing $(A.125) \times \hat{h}_2 - (A.129) \times (\hat{h}_1 - b_0)$, we obtain the following equation for C:

$$\hat{h}_3 C(\hat{h}_1 - b_0)^2 - (\hat{h}_2^2 + \hat{h}_4 C)(\hat{h}_1 - b_0) + \hat{h}_2 \hat{h}_3 C = 0, \qquad \text{(A.130)}$$

namely

$$(\hat{h}_2\hat{h}_3 - \hat{h}_4(\hat{h}_1 - b_0))C + \hat{h}_3(\hat{h}_1 - b_0)^2 - \hat{h}_2^2(\hat{h}_1 - b_0) = 0. \qquad \text{(A.131)}$$

The solution is

$$C = \frac{\hat{h}_3(\hat{h}_1 - b_0)^2 - \hat{h}_2^2(\hat{h}_1 - b_0)}{\hat{h}_4(\hat{h}_1 - b_0) - \hat{h}_2\hat{h}_3} = \frac{\hat{h}_3 Y^2 - \hat{h}_2^2 Y}{\hat{h}_4 Y - \hat{h}_2\hat{h}_3}, \qquad \text{(A.132)}$$

where we set

$$Y = (\hat{h}_1 - b_0). \qquad \text{(A.133)}$$

The equation (A.129) is written as

$$\hat{h}_2 Y^2 - 2\hat{h}_3 CY - \hat{h}_2^2 C + 2\hat{h}_4 C^2 = 0. \qquad \text{(A.134)}$$

Using (A.132), we obtain:

$$\hat{h}_2(\hat{h}_4 Y - \hat{h}_2 \hat{h}_3)^2 Y^2 - 2\hat{h}_3(\hat{h}_3 Y^2 - \hat{h}_2^2 Y)(\hat{h}_4 Y - \hat{h}_2 \hat{h}_3)Y$$
$$-\hat{h}_2^2(\hat{h}_3 Y^2 - \hat{h}_2^2 Y)(\hat{h}_4 Y - \hat{h}_2 \hat{h}_3)C + 2\hat{h}_4(\hat{h}_3 Y^2 - \hat{h}_2^2 Y)^2 = 0, \quad \text{(A.135)}$$

and using (A.132), we get

$$\hat{h}_2^3 \hat{h}_3^2 Y^4 + (2\hat{h}_2 \hat{h}_3^3 - 2\hat{h}_2^2 \hat{h}_3 - 3\hat{h}_2^2 \hat{h}_3 \hat{h}_4)Y^3 + 3\hat{h}_2^4 \hat{h}_4 Y^2 - \hat{h}_2^5 \hat{h}_3 Y = 0. \quad \text{(A.136)}$$

We know from (A.132) that $Y \neq 0$ because $C > 0$. Therefore we obtain the following equation for Y:

$$\hat{h}_2^3 \hat{h}_3^2 Y^3 + (2\hat{h}_2 \hat{h}_3^3 - 2\hat{h}_2^2 \hat{h}_3 - 3\hat{h}_2^2 \hat{h}_3 \hat{h}_4)Y^2 + 3\hat{h}_2^4 \hat{h}_4 Y - \hat{h}_2^5 \hat{h}_3 = 0. \quad \text{(A.137)}$$

From the discussions above, we obtain the following procedure for estimation.

Result:

Let Y be a solution of (A.137), and set

$$\hat{b}_0 = \hat{h}_1 - Y \qquad \text{(A.138)}$$

$$\hat{C} = \frac{\hat{h}_3 Y^2 - \hat{h}_2^2 Y}{\hat{h}_4 Y - \hat{h}_2 \hat{h}_3}, \qquad \text{(A.139)}$$

where we assume that $\hat{C}$ is positive. Next let $(\hat{x}_1, \hat{x}_2)$ be the solution of (A.122), where we suppose that $\hat{x}_1 < 0 < \hat{x}_2$, and set

$$\hat{c}_1 = -\frac{1}{\hat{x}_1} \qquad \text{(A.140)}$$

$$\hat{c}_2 = \frac{1}{\hat{x}_2}. \qquad \text{(A.141)}$$

These $(\hat{C}, \hat{c}_1, \hat{c}_2, \hat{b}_0)$ are estimators of the parameters (C, c_1, c_2, b_0) of a variance gamma process by the classical method of moments.

A.2.4.2 *Estimation by generalized method of moments*

We can apply also the generalized method of moments to the variance gamma model.

Estimation of C:

It is easy to see from the characteristic function $\phi_{VG}(u)$ given in (A.101) or (A.102) that

$$\lim_{u \to \infty} \frac{\text{Re}[\log \phi(u)]}{\log u} = -2C. \qquad \text{(A.142)}$$

Using this formula, we obtain an estimator $\hat{C}$ of C:

$$\hat{C} = -\frac{1}{2}\frac{\operatorname{Re}[\log\hat{\phi}(u_1)]}{u_1}, \tag{A.143}$$

for some large number u_1.

Estimation of b_0:

It is easy to see that

$$\lim_{u\to\infty}\frac{\operatorname{Im}[\log\phi(u)]}{u} = b_0. \tag{A.144}$$

Using this formula, we obtain an estimator $\hat{b}_0$ of b_0:

$$\hat{b}_0 = \frac{\operatorname{Im}[\log\hat{\phi}(u_2)]}{u_2}, \tag{A.145}$$

for some large number u_2.

Estimation of c_1 and c_2:

Let $\hat{C}$ and $\hat{b}_0$ be the estimators of C and b_0 respectively. The transformed moment-equations are given in this case as follows:

$$\hat{b}_0 + \hat{C}\left(\frac{1}{c_2} - \frac{1}{c_1}\right) = \hat{h}_1 \tag{A.146}$$

$$\hat{C}\left(\frac{1}{c_2^2} + \frac{1}{c_1^2}\right) = \hat{h}_2. \tag{A.147}$$

We can solve this equation for c_1 and c_2, and we obtain estimators $\hat{c}_1$ and $\hat{c}_2$.

A.2.5 *CGMY process*

The parameters of the CGMY process are (C, G, M, Y, b_1), and its Lévy measure is given by

$$\nu(dx) = C\left(I_{\{x<0\}}\exp(-G|x|) + I_{\{x>0\}}\exp(-M|x|)\right)|x|^{-(1+Y)}dx, \tag{A.148}$$

where $C > 0, G \geq 0, M \geq 0, Y < 2$. If $Y \leq 0$, then the conditions that $G > 0$ and $M > 0$ are assumed. We mention here that the case of $Y = 0$ is the variance gamma process case, and the case of $G = M = 0$ and $0 < Y < 2$ is the symmetric stable process case. In the sequel we assume that $G, M > 0$.

Since the condition

$$\int_{|x|\geq 1}|x|\nu(dx) < \infty \tag{A.149}$$

is satisfied, we can adopt the expression of the generating triplet in the form of $(0, \nu, b_1)_1$. If $Y < 1$, then the condition

$$\int_{|x|<1} |x|\nu(dx) < \infty \tag{A.150}$$

is satisfied. So in this case we have another expression of the generating triplet such that $(0, \nu, b_0)_0$.

A.2.5.1   The characteristic function $(\phi_{CGMY}(u))$

We first study the characteristic function of CGMY process.

(1) The case of $Y = 0$.

This case is variance gamma case. The characteristic function is

$$\phi_{VG}(u) = \exp\left\{ ib_0 u + \int_{|x|>0} (e^{iux} - 1)\nu(dx) \right\}$$

$$= \exp\left\{ ib_0 u - C\left(\log(1 + \frac{iu}{G}) + \log(1 - \frac{iu}{M}) \right) \right\}$$

$$= e^{ib_0 u} \left(\frac{1}{\left(1 + \frac{iu}{G}\right)\left(1 - \frac{iu}{M}\right)} \right)^C. \tag{A.151}$$

Remark A.3. If we adopt the expression $(0, \nu, b_1)_1$ of the generating triplet, then b_1 is given by

$$b_1 = b_0 + \int_{-\infty}^{\infty} x\nu(dx) = b_0 + C\left(\frac{1}{M} - \frac{1}{G} \right), \tag{A.152}$$

and

$$\phi_{VG}(u) = \exp\left\{ i\left(b_1 - C\left(\frac{1}{M} - \frac{1}{G} \right) \right) u \right.$$

$$\left. - C\left(\log(1 + \frac{iu}{G}) + \log(1 - \frac{iu}{M}) \right) \right\}$$

$$= e^{i\left(b_1 - C\left(\frac{1}{M} - \frac{1}{G}\right)\right)u} \left(\frac{1}{\left(1 + \frac{iu}{G}\right)\left(1 - \frac{iu}{M}\right)} \right)^C. \tag{A.153}$$

(2) The case of $Y < 1$ and $Y \neq 0$.

In this case we obtain:

$$\phi_{CGMY}(u) = \exp\left\{ ib_0 u + \int_{|x|>0} (e^{iux} - 1)\nu(dx) \right\}$$

$$= \exp\left\{ ib_0 u + C\Gamma(-Y) \left((M - iu)^Y - M^Y + (G + iu)^Y - G^Y \right) \right\}. \tag{A.154}$$

If we adopt another expression of the generating triplet, $(0, \nu(dx), b_1)_1$, then we obtain:

$$
\begin{aligned}
\phi_{CGMY}(u) &= \exp\left\{ ib_1 u + \int_{|x|>0} (e^{iux} - 1 - iux)\nu(dx) \right\} \\
&= \exp\left\{ i\left(b_1 + C\Gamma(-Y)Y(M^{Y-1} - G^{Y-1})\right) u \right. \\
&\quad \left. + C\Gamma(-Y)\left((M - iu)^Y - M^Y + (G + iu)^Y - G^Y\right)\right\}.
\end{aligned}
$$

$$(A.155)$$

Remark A.4.

(i) It follows from these formulas that

$$
b_0 = b_1 + C\Gamma(-Y)Y(M^{Y-1} - G^{Y-1}). \tag{A.156}
$$

(ii) The right-hand side of (A.154) or (A.155) converged to the characteristic function of variance gamma when $Y \to 0$, namely:

$$
\begin{aligned}
\lim_{Y \to 0} &\left\{ ib_0 u - C\frac{\Gamma(1-Y)}{Y}\left((M - iu)^Y - M^Y + (G + iu)^Y - G^Y\right)\right\} \\
&= ib_0 u - C\left(\log(M - iu) - \log M + \log(G + iu) - \log G\right) \\
&= ib_0 u - C\left(\log(1 - \frac{iu}{M}) + \log(1 + \frac{iu}{G})\right).
\end{aligned}
$$

$$(A.157)$$

(3) The case of $Y = 1$.

In this case we obtain:

$$
\begin{aligned}
\phi_{CGMY}(u) &= \exp\left\{ ib_1 u + \int_{|x|>0} (e^{iux} - 1 - iux)\nu(dx) \right\} \\
&= \exp\left\{ ib_1 u + C\left(\left((M - iu)\log(1 - \frac{iu}{M}) + iu\right)\right.\right. \\
&\quad \left.\left. + \left((G + iu)\log(1 + \frac{iu}{G}) - iu\right)\right)\right\} \\
&= \exp\left\{ ib_1 u + C\left((M - iu)\log(1 - \frac{iu}{M})\right.\right. \\
&\quad \left.\left. + (G + iu)\log(1 + \frac{iu}{G})\right)\right\}.
\end{aligned}
$$

$$(A.158)$$

(4) The case of $1 < Y < 2$**.**

$$\phi_{CGMY}(u) = \exp\left\{ ib_1 u + \int_{|x|>0} (e^{iux} - 1 - iux)\nu(dx) \right\}$$

$$= \exp\left\{ ib_1 u + C\Gamma(-Y)\left((M - iu)^Y - M^Y + (G + iu)^Y - G^Y\right) \right.$$
$$\left. + iuC\Gamma(-Y)Y\left(M^{Y-1} - G^{Y-1}\right) \right\}$$

$$= \exp\left\{ i\left(b_1 + C\Gamma(-Y)Y\left(M^{Y-1} - G^{Y-1}\right)\right)u \right.$$
$$\left. + C\Gamma(-Y)\left((M - iu)^Y - M^Y + (G + iu)^Y - G^Y\right) \right\}. \quad (A.159)$$

Remark A.5. The last formula is same with case (2).

Remark A.6. The right-hand side of (A.159) converges to the right-hand side of (A.158), namely:

$$\lim_{Y \to 1} \left\{ i\left(b_1 + C\Gamma(-Y)Y\left(M^{Y-1} - G^{Y-1}\right)\right)u \right.$$
$$\left. + C\Gamma(-Y)\left((M - iu)^Y - M^Y + (G + iu)^Y - G^Y\right) \right\}$$

$$= \lim_{Y \to 1} \left\{ i\left(b_1 + C\frac{\Gamma(2 - Y)}{Y - 1}\left(M^{Y-1} - G^{Y-1}\right)\right)u \right.$$
$$+ C\frac{\Gamma(2 - Y)}{Y(Y - 1)}\left((M - iu)((M - iu)^{Y-1} - 1) - M(M^{Y-1} - 1)\right.$$
$$\left.\left. + (G + iu)((G + iu)^{Y-1} - 1) - G(G^{Y-1} - 1)\right) \right\}$$

$$= i\left(b_1 + C\left(\log M - \log G\right)\right)u$$
$$+ C\left((M - iu)\log(M - iu) - M\log M + (G + iu)\log(G + iu) - G\log G\right)$$
$$= ib_1 u + C\left((M - iu)\log(1 - \frac{iu}{M}) + (G + iu)\log(1 + \frac{iu}{G})\right). \quad (A.160)$$

A.2.5.2 *Properties of the characteristic function*

We next study the properties of the characteristic function of the CGMY process.

Proposition A.3. *It holds that*

$$\lim_{u \to \infty} \frac{\log |\mathrm{Re}[\log \phi_{CGMY}(u)]|}{\log u} = \begin{cases} Y, & Y > 0, \\ 0, & Y \le 0. \end{cases} \quad (A.161)$$

(Proof (See Section B.2.5.4.))

Proposition A.4. (i) *If* $0 < Y < 2$, *then it holds that*

$$\lim_{u \to \infty} \frac{\mathrm{Re}[\log \phi_{CGMY}(u)]}{u^Y} = \begin{cases} 2C\Gamma(-Y)\cos\left(\frac{\pi}{2}Y\right), & Y \ne 1 \\ -\pi C, & Y = 1. \end{cases} \quad (A.162)$$

(ii) *If Y = 0, then it holds that*

$$\lim_{u \to \infty} \frac{\text{Re}[\log \phi_{CGMY}(u)]}{\log u} = -2C. \tag{A.163}$$

(iii) *If Y < 0*

$$\lim_{u \to \infty} \text{Re}[\log \phi_{CGMY}(u)] = -C\Gamma(-Y)(M^Y + G^Y). \tag{A.164}$$

(Proof (See Section B.2.5.5.))

Remark A.7. It holds that

$$\lim_{Y \to 1} 2C\Gamma(-Y) \cos\left(\frac{\pi}{2}Y\right) = -\pi C. \tag{A.165}$$

Set the following:

$$\tilde{b}_0 = \begin{cases} b_1 + C\Gamma(-Y)Y \left(M^{Y-1} - G^{Y-1}\right), & Y \neq 0, 1 \\ b_1 - C\left(\frac{1}{M} - \frac{1}{G}\right) (= b_0), & Y = 0 \\ b_1 + C\left(\log M - \log G\right), & Y = 1. \end{cases} \tag{A.166}$$

(If $Y < 1$, then $\tilde{b}_0$ is identical with b_0 of $(\sigma^2, \nu, b_0)_0$.)
This leads to the following.

Proposition A.5. *It holds that*

$$\lim_{u \to \infty} \frac{\text{Im}[\log \phi_{CGMY}(u)]}{u} = \tilde{b}_0. \tag{A.167}$$

(Proof (See Section B.2.5.6.))

A.2.5.3 *Estimation*

Since CGMY processes have moments of any order, we can apply the classical method of moments to the estimation problem of the CGMY process. However, the equations obtained in the classical method of moments are not easy to solve. So we try to combine the generalized method of moments (= characteristic function method) with the classical method of moments.

Judgment of $Y > 0$ or $Y \le 0$:
We can use the result of Proposition A.3 to judge whether $Y > 0$ or $Y \le 0$. If the value

$$\frac{\log |\text{Re}[\log \hat{\phi}_{CGMY}(u)]|}{\log u} \tag{A.168}$$

is positive for large u, we judge as $Y > 0$. If not, we judge as $Y \le 0$.

(1) Case of $Y > 0$.

Assume that we have judged that $Y > 0$. Then we can continue the following procedure.

(i) Estimation of the value of Y:

Using Proposition A.3 again, we obtain an estimator $\hat{Y}$ of Y:

$$\hat{Y} = \frac{\log |\text{Re}[\log \hat{\phi}_{\text{CGMY}}(u_1)]|}{\log u_1}, \tag{A.169}$$

where u_1 is a number large enough.

(ii) Estimation of b_1:

It is known that $b_1 = m_1$ (=mean). Therefore we obtain an estimator:

$$\hat{b}_1 = \hat{h}_1 \quad (= \hat{m}_1). \tag{A.170}$$

(iii) Estimation of $\tilde{b}_0$:

From the result of Proposition A.5, we obtain an estimator $\hat{\tilde{b}}_0$ of $\tilde{b}_0$:

$$\hat{\tilde{b}}_0 = \frac{\text{Im}[\log \hat{\phi}_{CGMY}(u_2)]}{u_2}, \tag{A.171}$$

for some large number u_2.

(iv) Estimation of C:

(a) The case $0 < Y < 2, Y \neq 1$:

From result (i) of Proposition A.4, we obtain an estimator $\hat{C}$ solving the following equation:

$$\frac{\text{Re}[\log \hat{\phi}_{CGMY}(u_3)]}{(u_3)^{\hat{Y}}} = 2C\Gamma(-\hat{Y}) \cos\left(\frac{\pi}{2}\hat{Y}\right), \tag{A.172}$$

for some large number u_3.

(b) The case of $Y = 1$:

From result (i) of Proposition A.4, we obtain an estimator $\hat{C}$ solving the following equation:

$$\frac{\text{Re}[\log \hat{\phi}_{CGMY}(u_3)]}{(u_3)^{\hat{Y}}} = -\pi C, \tag{A.173}$$

for some large number u_3.

(v) Estimation of G and M:

Let $\hat{Y}$, $\hat{b}_1$, $\hat{\tilde{b}}_0$ and $\hat{C}$ be estimators of Y, b_1, $\tilde{b}_0$ and C respectively.

(a) The case $0 < Y < 2, Y \neq 1$:

The transformed moment-equations (A.29) and (A.30) are in this case:

$$\hat{\bar{b}}_0 + \hat{C}\Gamma(-\hat{Y})\hat{Y}\left(G^{\hat{Y}-1} - M^{\hat{Y}-1}\right) = \hat{h}_1 = \hat{b}_1, \qquad (A.174)$$

$$\hat{C}\Gamma(-\hat{Y})\hat{Y}(\hat{Y}-1)\left(G^{\hat{Y}-2} + M^{\hat{Y}-2}\right) = \hat{h}_2. \qquad (A.175)$$

We can solve this equation for G and M, and obtain the estimators $\hat{G}$ and $\hat{M}$.

(b) The case of $Y = 1$:

The moment-equations (A.29) and (A.30) in this case are

$$\hat{\bar{b}}_0 + \hat{C}\left(\log G - \log M\right) = \hat{h}_1 = \hat{b}_1, \qquad (A.176)$$

$$-\hat{C}\left(\frac{1}{M} + \frac{1}{G}\right) = \hat{h}_2. \qquad (A.177)$$

We can solve this equation for G and M, and obtain estimators $\hat{G}$ and $\hat{M}$.

(2) Case of $Y \leq 0$.

Judgment of $Y = 0$ or $Y < 0$.

When we have judged as $Y \leq 0$, then we have to check whether $Y = 0$ or $Y < 0$. This is done by the use of the results of (ii) and (iii) of Proposition A.4.

If the limit

$$\lim_{u\to\infty} \mathrm{Re}[\log \phi_{CGMY}(u)] \qquad (A.178)$$

diverges to $-\infty$, then we judge it as $Y = 0$. If it converges, then we judge it as $Y < 0$.

(3) Case of $Y = 0$.

(i) Estimation of C:

Using the following formula:

$$\lim_{u\to\infty} \frac{\mathrm{Re}[\log \phi_{CGMY}(u)]}{\log u} = -2C, \qquad (A.179)$$

we obtain an estimator $\hat{C}$ of C:

$$\hat{C} = -\frac{1}{2}\frac{\mathrm{Re}[\log \hat{\phi}_{CGMY}(u_1)]}{u_1}, \qquad (A.180)$$

for some large number u_1.

(ii) Estimation of b_0:

It is easy to see that

$$\lim_{u \to \infty} \frac{\mathrm{Im}[\log \phi_{CGMY}(u)]}{u} = b_0. \tag{A.181}$$

Using this formula, we obtain an estimator $\hat{b}_0$ of b_0:

$$\hat{b}_0 = \frac{\mathrm{Im}[\log \hat{\phi}_{CGMY}(u_2)]}{u_2}, \tag{A.182}$$

for some large number u_2.

(iii) Estimation of G and M:

Let $\hat{C}$ and $\hat{b}_0$ be the estimators of C and b_0 respectively. The transformed moment-equations (A.29) and (A.30) in this case are:

$$\hat{b}_0 + \hat{C}\left(\frac{1}{M} - \frac{1}{G}\right) = \hat{h}_1, \tag{A.183}$$

$$\hat{C}\left(\frac{1}{M^2} + \frac{1}{G^2}\right) = \hat{h}_2. \tag{A.184}$$

We can solve this equation for G and M, and we obtain estimators $\hat{G}$ and $\hat{M}$ of G and M.

(4) Case of $Y < 0$.

(i) Estimation of b_1:

An estimator $\hat{b}_1$ is given by

$$\hat{b}_1 = \hat{h}_1 \quad (= \hat{m}_1). \tag{A.185}$$

(ii) Estimation of $\tilde{b}_0$:

From the result of Proposition A.5, we obtain the estimator $\hat{\tilde{b}}_0$ of $\tilde{b}_0$:

$$\hat{\tilde{b}}_0 = \frac{\mathrm{Im}[\log \phi_{CGMY}(u_2)]}{u_2}, \tag{A.186}$$

for some large number u_2.

(iii) Estimation of Y, C, G and M:

Let $\hat{b}_1$ and $\hat{\tilde{b}}_0$ be the estimators of b_1 and $\tilde{b}_0$. By (iii) of Proposition A.4, we obtain the equation:

$$\mathrm{Re}[\log \hat{\phi}_{CGMY}(u_3)] = -C\Gamma(-Y)(M^Y + G^Y), \tag{A.187}$$

for some large u_3. The transformed moment-equations (A.29), (A.30), and (A.31) in this case are:

$$\hat{\tilde{b}}_0 + C\Gamma(-Y)Y\left(G^{Y-1} - M^{Y-1}\right) = \hat{h}_1, \tag{A.188}$$

$$C\Gamma(-Y)Y(Y-1)\left(G^{Y-2} + M^{Y-2}\right) = \hat{h}_2, \tag{A.189}$$

$$C\Gamma(-Y)Y(Y-1)(Y-2)\left(G^{Y-3} - M^{Y-3}\right) = \hat{h}_3. \tag{A.190}$$

Solving the above four equations for Y, C, G, M, we obtain estimators $\hat{Y}, \hat{C}, \hat{G}$, and $\hat{M}$.

A.2.5.4 *Proof of Proposition A.3*

(Proof)

(1) The case of $0 < Y < 2$, $Y \neq 1$:

The characteristic function is

$$\phi_{CGMY}(u)$$
$$= \exp \left\{ i \left(b_1 + C\Gamma(-Y)Y \left(M^{Y-1} - G^{Y-1} \right) \right) u \right.$$
$$\left. + C\Gamma(-Y) \left((M - iu)^Y - M^Y + (G + iu)^Y - G^Y \right) \right\}. \quad (A.191)$$

Set the following:

$$M - iu = \left(\sqrt{M^2 + u^2} \right) e^{i\theta_M(u)}, \quad (A.192)$$

$$G + iu = \left(\sqrt{G^2 + u^2} \right) e^{i\theta_G(u)}. \quad (A.193)$$

Then we obtain:

$$\mathrm{Re}[\log \phi_{CGMY}(u)]$$
$$= C\Gamma(-Y) \left(\left(\sqrt{M^2 + u^2} \right)^Y \cos\left(Y\theta_M(u) \right) - M^Y \right.$$
$$\left. + \left(\sqrt{G^2 + u^2} \right)^Y \cos\left(Y\theta_G(u) \right) - G^Y \right). \quad (A.194)$$

We remark that $\theta_M(u)$ is a decreasing function and $\theta_G(u)$ is an increasing function, and that

$$\lim_{u \to \infty} \theta_M(u) = -\frac{\pi}{2}, \quad (A.195)$$

$$\lim_{u \to \infty} \theta_G(u) = \frac{\pi}{2}. \quad (A.196)$$

Therefore it holds that

$$\lim_{u \to \infty} \cos\left(Y\theta_M(u) \right) = \cos\left(-\frac{\pi}{2}Y \right) = \cos\left(\frac{\pi}{2}Y \right), \quad (A.197)$$

$$\lim_{u \to \infty} \cos\left(Y\theta_G(u) \right) = \cos\left(\frac{\pi}{2}Y \right). \quad (A.198)$$

If $0 < Y < 1$, then $\cos\left(\frac{\pi}{2}Y \right)$ is positive. If $1 < Y < 2$, then $\cos\left(\frac{\pi}{2}Y \right)$ is negative. From these facts it follows that $|\mathrm{Re}[\log \phi_{CGMY}(u)]|$ is of order $|u|^Y$ when $u \to \infty$. So we have obtained the result (A.161).

(2) The case of $Y = 1$:

In this case, the characteristic function is

$$\phi_{CGMY}(u)$$
$$= \exp \left\{ ib_1 u + C \left((M - iu) \log(1 - \frac{iu}{M}) + (G + iu) \log(1 + \frac{iu}{G}) \right) \right\}. $$
$$(A.199)$$

Using the same notations as above, we obtain:

$$\log \phi_{CGMY}(u)$$

$$= ib_1 u + C \left(\sqrt{M^2 + u^2} e^{i\theta_M(u)} \left(\log \sqrt{1 + \frac{u^2}{M^2}} + i\theta_M(u) \right) \right.$$

$$\left. + \sqrt{G^2 + u^2} e^{i\theta_G(u)} \left(\log \sqrt{1 + \frac{u^2}{G^2}} + i\theta_G(u) \right) \right), \qquad (A.200)$$

and

$$\mathrm{Re}[\log \phi_{CGMY}(u)]$$

$$= C \left(\sqrt{M^2 + u^2} \left(\log \sqrt{1 + \frac{u^2}{M^2}} \cos \theta_M(u) - \theta_M(u) \sin \theta_M(u) \right) \right.$$

$$\left. + \sqrt{G^2 + u^2} \left(\log \sqrt{1 + \frac{u^2}{G^2}} \cos \theta_G(u) - \theta_G(u) \sin \theta_G(u) \right) \right)$$

$$= C \left(M \log \sqrt{1 + \frac{u^2}{M^2}} + \theta_M(u) u + G \log \sqrt{1 + \frac{u^2}{G^2}} - \theta_G(u) u \right), \quad (A.201)$$

where we use

$$\cos \theta_M(u) = \frac{M}{\sqrt{M^2 + u^2}}, \quad \sin \theta_M(u) = \frac{-u}{\sqrt{M^2 + u^2}}, \qquad (A.202)$$

$$\cos \theta_G(u) = \frac{G}{\sqrt{G^2 + u^2}}, \quad \sin \theta_G(u) = \frac{u}{\sqrt{G^2 + u^2}}. \qquad (A.203)$$

The formula (A.161) follows from (A.201).

(3) The case of $Y = 0$:

In this case,

$$\phi_{CGMY}(u) = \phi_{VG}(u)$$

$$= \exp \left\{ ib_0 u - C \left(\log(1 + \frac{iu}{G}) + \log(1 - \frac{iu}{M}) \right) \right\}. \qquad (A.204)$$

Therefore:

$$\mathrm{Re}[\log \phi_{CGMY}(u)]$$

$$= -C \left(\log \sqrt{1 + \frac{u^2}{M^2}} + \log \sqrt{1 + \frac{u^2}{G^2}} \right). \qquad (A.205)$$

The formula (A.161) follows from (A.205).

(4) The case of $Y < 0$:

In this case,

$$\phi_{CGMY}(u)$$
$$= \exp\left\{ib_0 u + C\Gamma(-Y)\left((M - iu)^Y - M^Y + (G + iu)^Y - G^Y\right)\right\},$$
(A.206)

and

$$\mathrm{Re}[\log \phi_{CGMY}(u)]$$
$$= C\Gamma(-Y)\left(\left(\sqrt{M^2 + u^2}\right)^Y \cos\left(Y\theta_M(u)\right) - M^Y\right.$$
$$\left. + \left(\sqrt{G^2 + u^2}\right)^Y \cos\left(Y\theta_G(u)\right) - G^Y\right).$$
(A.207)

Since $Y < 0$, it is easy to see that the formula (A.161) holds true.
(Q.E.D)

A.2.5.5 *Proof of Proposition A.4*

(Proof)

(1) The case of $0 < Y < 2$, $Y \neq 1$:
It holds that

$$\mathrm{Re}[\log \phi_{CGMY}(u)]$$
$$= C\Gamma(-Y)\left(\left(\sqrt{M^2 + u^2}\right)^Y \cos\left(Y\theta_M(u)\right) - M^Y\right.$$
$$\left. + \left(\sqrt{G^2 + u^2}\right)^Y \cos\left(Y\theta_G(u)\right) - G^Y\right),$$
(A.208)

and

$$\lim_{u \to \infty} \cos\left(Y\theta_M(u)\right) = \cos\left(-\frac{\pi}{2}Y\right) = \cos\left(\frac{\pi}{2}Y\right),$$
(A.209)

$$\lim_{u \to \infty} \cos\left(Y\theta_G(u)\right) = \cos\left(\frac{\pi}{2}Y\right).$$
(A.210)

Using these formulas, we obtain:

$$\lim_{u \to \infty} \frac{\mathrm{Re}[\log \phi_{CGMY}(u)]}{u^Y} = 2C\Gamma(-Y)\cos\left(\frac{\pi}{2}Y\right).$$
(A.211)

(2) The case $Y = 1$:

$$\mathrm{Re}[\log \phi_{CGMY}(u)] = C\left(M\log\sqrt{1 + \frac{u^2}{M^2}} + \theta_M(u)u\right.$$
$$\left. + G\log\sqrt{1 + \frac{u^2}{G^2}} - \theta_G(u)u\right),$$
(A.212)

and

$$\lim_{u\to\infty} \frac{\mathrm{Re}[\log \phi_{CGMY}(u)]}{u} = -\pi C. \tag{A.213}$$

(3) The case of $Y = 0$:
In this case, it holds that

$$\mathrm{Re}[\log \phi_{\mathrm{CGMY}}(\mathrm{u})] = -C \left(\log \sqrt{1 + \frac{u^2}{M^2}} + \log \sqrt{1 + \frac{u^2}{G^2}} \right), \tag{A.214}$$

and so

$$\lim_{u\to\infty} \frac{\mathrm{Re}[\log \phi_{CGMY}(u)]}{\log u} = -2C. \tag{A.215}$$

(4) The case $Y < 0$:
It holds that

$$\mathrm{Re}[\log \phi_{CGMY}(u)] = C\Gamma(-Y) \left(\left(\sqrt{M^2 + u^2}\right)^Y \cos\left(Y\theta_M(u)\right) - M^Y \right.$$
$$\left. + \left(\sqrt{G^2 + u^2}\right)^Y \cos\left(Y\theta_G(u)\right) - G^Y \right). \tag{A.216}$$

Since $Y < 0$, from this formula we obtain:

$$\lim_{u\to\infty} \mathrm{Re}[\log \phi_{CGMY}(u)] = -C\Gamma(-Y)(M^Y + G^Y). \tag{A.217}$$

(Q.E.D.)

A.2.5.6   *Proof of Proposition A.5*

(Proof)

(1) The case of $Y \neq 0, 1$:
It holds that

$$\phi_{CGMY}(u) = \exp\left\{i\tilde{b}_0 u + C\Gamma(-Y)\left((M - iu)^Y - M^Y \right.\right.$$
$$\left.\left. + (G + iu)^Y - G^Y\right)\right\}, \tag{A.218}$$

$$\mathrm{Im}[\log \phi_{CGMY}(u)] = \tilde{b}_0 u + C\Gamma(-Y) \left(\left(\sqrt{M^2 + u^2}\right)^Y \sin\left(Y\theta_M(u)\right) \right.$$
$$\left. + \left(\sqrt{G^2 + u^2}\right)^Y \sin\left(Y\theta_G(u)\right) \right), \tag{A.219}$$

and

$$\lim_{u \to \infty} \sin\left(Y\theta_M(u)\right) = \sin\left(-\frac{\pi}{2}Y\right), \tag{A.220}$$

$$\lim_{u \to \infty} \sin\left(Y\theta_G(u)\right) = \sin\left(\frac{\pi}{2}Y\right). \tag{A.221}$$

It should be noted here that $\sin\left(-\frac{\pi}{2}Y\right) = -\sin\left(\frac{\pi}{2}Y\right)$.

If $Y < 1, Y \neq 0$, then it is easy to check the formula (A.167). So we assume that $1 < Y < 2$.

The following is a trivial inequality.

$$\left| \left(\sqrt{M^2 + u^2}\right)^Y \sin\left(Y\theta_M(u)\right) + \left(\sqrt{G^2 + u^2}\right)^Y \sin\left(Y\theta_G(u)\right) \right|$$

$$\leq \left| \left(\left(\sqrt{G^2 + u^2}\right)^Y - \left(\sqrt{M^2 + u^2}\right)^Y \right) \sin\left(Y\theta_G(u)\right) \right|$$

$$+ \left| \left(\sqrt{M^2 + u^2}\right)^Y \left(\sin\left(Y\theta_G(u)\right) + \sin\left(Y\theta_M(u)\right)\right) \right|. \tag{A.222}$$

The first term of the right-hand side of (A.222) satisfies the following inequality:

$$\left| \left(\left(\sqrt{G^2 + u^2}\right)^Y - \left(\sqrt{M^2 + u^2}\right)^Y \right) \sin\left(Y\theta_G(u)\right) \right|$$

$$\leq |u|^Y \left| \left(1 + \frac{G^2}{u^2}\right)^{\frac{Y}{2}} - \left(1 + \frac{M^2}{u^2}\right)^{\frac{Y}{2}} \right|$$

$$\leq |u|^Y \left| \frac{G^2}{u^2} - \frac{M^2}{u^2} \right|, \quad \text{for large } u$$

$$\leq |G^2 - M^2||u|^{Y-2}, \quad \text{for large } u, \tag{A.223}$$

where we have used the assumption that $Y < 2$.

The second term of the right-hand side of (A.222) satisfies the following inequality:

$$\left| \left(\sqrt{M^2 + u^2} \right)^Y \left(\sin\left(Y\theta_G(u) \right) + \sin\left(Y\theta_M(u) \right) \right) \right|$$

$$\leq 2 \left(\sqrt{M^2 + u^2} \right)^Y \left| \sin\left(\frac{Y(\theta_G(u) + \theta_M(u))}{2} \right) \cos\left(\frac{Y(\theta_G(u) - \theta_M(u))}{2} \right) \right|$$

$$\leq 2 \left(\sqrt{M^2 + u^2} \right)^Y \left| \sin\left(\frac{Y(\theta_G(u) + \theta_M(u))}{2} \right) \right|$$

$$\leq 2 \left(\sqrt{M^2 + u^2} \right)^Y \left| \sin\left((\theta_G(u) + \theta_M(u)) \right) \right|, \quad \text{since } Y < 2$$

$$\leq 2 \left(\sqrt{M^2 + u^2} \right)^Y \left| \sin\left(\theta_G(u) \right) \cos\left(\theta_M(u) \right) + \cos\left(\theta_G(u) \right) \sin\left(\theta_M(u) \right) \right|$$

$$\leq 2 \left(\sqrt{M^2 + u^2} \right)^Y \left| \frac{u}{\sqrt{G^2 + u^2}} \frac{M}{\sqrt{M^2 + u^2}} - \frac{G}{\sqrt{G^2 + u^2}} \frac{u}{\sqrt{M^2 + u^2}} \right|$$

$$= 2 \left(\sqrt{M^2 + u^2} \right)^{Y-1} \left| \frac{|M - G|u}{\sqrt{G^2 + u^2}} \right|. \tag{A.224}$$

Since $Y < 2$, from (A.219), (A.222), (A.223), and (A.224), the result (A.167) follows.

(2) The case of $Y = 0$:
The result is trivial.

(3) The case of $Y = 1$:
It holds that

$$\phi_{CGMY}(u)$$

$$= \exp\left\{ ib_1 u + C \left((M - iu)\log(1 - \frac{iu}{M}) + (G + iu)\log(1 + \frac{iu}{G}) \right) \right\}$$

$$= \exp\left\{ i\left(b_1 + C(\log M - \log G) \right)u \right.$$

$$+ C\left((M - iu)\log(M - iu) - M\log M \right.$$

$$\left. + (G + iu)\log(G + iu) - G\log G \right) \right\}, \tag{A.225}$$

and

$$\text{Im}[\log \phi_{CGMY}(u)]$$

$$= \tilde{b}_1 u + C \left(\sqrt{M^2 + u^2} \left(\log \sqrt{M^2 + u^2} \sin \theta_M(u) + \theta_M(u) \cos \theta_M(u) \right) \right.$$

$$\left. + \sqrt{G^2 + u^2} \left(\log \sqrt{G^2 + u^2} \sin \theta_G(u) + \theta_G(u) \cos \theta_G(u) \right) \right)$$

$$= \tilde{b}_1 u + C \left(-u \log \sqrt{M^2 + u^2} + M\theta_M(u) + u \log \sqrt{G^2 + u^2} + G\theta_G(u) \right). \tag{A.226}$$

It is easy to see that the following formula holds true:

$$\left| -u \log \sqrt{M^2 + u^2} + u \log \sqrt{G^2 + u^2} \right|$$
$$= |u| \left| \log \sqrt{G^2 + u^2} - \log \sqrt{M^2 + u^2} \right|$$
$$= \frac{1}{2}|u| \left| \log \left(\frac{G^2 + u^2}{M^2 + u^2} \right) \right|. \tag{A.227}$$

From the above three formulas, we obtain the following result:

$$\lim_{u \to \infty} \frac{\mathrm{Im}[\log \phi_{CGMY}(u)]}{u}$$
$$= \tilde{b}_0 + \lim_{u \to \infty} \frac{\frac{1}{2}|u| \left| \log \left(\frac{G^2 + u^2}{M^2 + u^2} \right) \right|}{u} = \tilde{b}_0. \tag{A.228}$$

(Q.E.D.)

Notes

This chapter is based on Miyahara (2002) [87].

The generalized method of moments is not necessarily best from the theoretical point of view, but it is very convenient from the practical point of view. In fact, it is very easy to apply this method to various kinds of estimation problems.

The moment estimation for stable processes was studied by Press (1972) [102]. Modifying his idea, we have developed the generalized method of moments for the estimation of stable processes and other Lévy processes.

For the further study of the estimation of Lévy processes, Kunitomo and Owada (2004) [72] is very useful.

Bibliography

[1] Applebaum, D. (2004), *Lévy Processes and Stochastic Calculus*, Cambridge University Press, Cambridge.

[2] Barndorff-Nielsen, O. E. (1995), Normal Inverse Gaussian Distributions and the Modeling of Stock Returns, *Research Report* 300, Department of Theoretical Statistics, Aarhus University.

[3] Barndorff-Nielsen, O. E. and Halgreen, O. (1977), Infinite Divisibility of the Hyperbolic and General Inverse Gaussian Distributions, *Z. fur Wahr. und ver. Gebiete* 38, 309–312.

[4] Barndorff-Nielsen, O. E., Mikosch, T. and Resnick, S. I. (eds) (2001), *Lévy Processes: Theory and Applications*, Birkhäuser, Berlin.

[5] Becherer, D. (2003), Rational Hedging and Valuation of Integrated Risks under Constant Absolute Risk Aversion, *Insurance: Mathematics and Economics* 33, 1–28.

[6] Bellini, F. and Frittelli, M. (2002), On the Existence of Minimax Martingale Measures, *Mathematical Finance* 12, 1–21.

[7] Bender, C. and Niethammer, C. (2008), On q-Optimal Martingale Measures in Exponential Lévy Models, *Finance and Stochastics* 12(3), 381–410.

[8] Bertoin, J. (1996), *Lévy Processes*, Cambridge University Press, Cambridge.

[9] Björk, T. (2004), *Arbitrage Theory in Continuous Time*, second edition, Oxford University Press, Oxford.

[10] Black, F. and Scholes, M. (1973), The Pricing of Options and Corporate Liabilities, *Journal of Political Economy* 81, 637–654.

[11] Bühlmann, H. (1970), *Mathematical Methods in Risk Theory*, Springer, Berlin.

[12] Bühlmann, H., Delbaen, F., Embrechts, P. et al. (1996), No-arbitrage, Change of Measure and Conditional Esscher Transforms, *CWI Quarterly* 9(4), 291–317.

[13] Carmona, R. (ed.) (2009), *Indifference Pricing: Theory and Applications*, Princeton University Press, Princeton.

[14] Carr, P., Geman, H., Madan, D.B. et al. (2002), The Fine Structure of Asset Returns: An Empirical Investigation, *Journal of Business* 75, 305–332.

[15] Carr, P. and Madan, D. (1999), Option Valuation Using the Fast Fourier transform, *Journal of Computational Finance* 2, 61–73.

[16] Carr, P. and Wu, L. (2003), The Finite Moment Log Stable Processes and Option Pricing, *The Journal of Finance* 58(2), 753–777.

[17] Carr, P. and Wu, L. (2004), Time-changed Lévy Processes and Option Pricing, *Journal of Financial Economics* 71, 113–141.

[18] Chan, T. (1999), Pricing Contingent Claims on Stocks Derived by Lévy Processes, *The Annals of Applied Probability* 9(2), 504–528.

[19] Cheang, G. H. L. (2004), A Simple Approach to Pricing Options with Jumps, (working paper).

[20] Cheridito, P., Delbaen, F. and Kupper, M. (2006), Dynamic Monetary Risk Measures for Bounded Discrete-Time Processes, *Electronic Journal of Probability* 11, 57–106.

[21] Cheridito, P. and Kupper, M. (2006), Time-Consistency of Indifference Prices and Monetary Utility Functions (preprint).

[22] Choulli, T. and Stricker, C. (2005), Minimal Entropy-Hellinger Martingale Measure in Incomplete Markets, *Mathematical Finance* 15, 465–490.

[23] Choulli, T. and Stricker, C. (2006), More on Minimal Entropy-Hellinger Martingale Measure, *Mathematical Finance* 16, 1–19.

[24] Choulli, T. and Stricker, C. (2007), Minimal Hellinger Martingale Measures of Order q, *Finance and Stochastics* 11, 399–427.

[25] Cont, R. and Tankov, P. (2004a), *Financial Modeling with Jump Processes,* Chapman & Hall/CRC, London.

[26] Cont, R. and Tankov, P. (2004b), Nonparametric Calibration of Jump-diffusion option Pricing Models, *Journal of Computational Finance* 7, 1–49.

[27] Cover, T. and Thomas, J. (1991), *Elements of Information Theory,* Wiley, Chichester.

[28] Csiszar, I. (1975), I-divergence Geometry of Probability Distributions and Minimization Problems, *Annals of Probability* 3(1), 146–158.

[29] Delbaen, F., Grandits, P., Rheinländer, T. et al. (2002), Exponential Hedging and Entropic Penalties, *Mathematical Finance* 12(2), 99–124.

[30] Delbaen, F. and Schachermayer, W. (2006), *The Mathematics of Arbitrage,* Springer, Berlin.

[31] Dixit, A. and Pindyck, R. S. (1994), *Investment under Uncertainty,* Princeton University Press, Princeton.

[32] Eberlein, E. and Keller, U. (1995), Hyperbolic Distributions in Finance, *Bernoulli* 1, 281–299.

[33] Edelman, D. (1995), A Note: Natural Generalization of Black–Scholes in the Presence of Skewness, Using Stable Processes, *ABACUS* 31(1), 113–119.

[34] Esche, F. and Schweizer, M. (2005), Minimal Entropy Preserves Lévy Property: How and Why, *Stochastic Processes and their Applications* 115, 299–327.

[35] Esscher, F. (1932), On the Probability Function in the Collective Theory of Risk, *Skandinavisk Aktuarietidskrift* 15, 175–195.

[36] Fama, E. F. (1963), Mandelbrot and the Stable Paretian Hypothesis, *Journal of Business* 36, 420–429.

[37] Feller, W. (1966), *An Introduction to Probability Theory and Its Applications*, Wiley, Chichester.

[38] Föllmer, H. and Schied, A. (2004), *Stochastic Finance*, second edition, Walter de Gruyter, Berlin and New York.

[39] Föllmer, H. and Schweizer, M. (1991), Hedging of Contingent Claims under Incomplete Information, in Davis, M.H.A. and Elliot, R.J. (eds), *Applied Stochastic Analysis*, Gordon and Breach, Newark, 389–414.

[40] Frittelli, M. (2000a), The Minimal Entropy Martingale Measures and the Valuation Problem in Incomplete Markets, *Mathematical Finance* 10, 39–52.

[41] Frittelli, M. (2000b), Introduction to a Theory of Value Coherent with the No Arbitrage Principle, *Finance and Stochastics* 4, 275–297.

[42] Fujisaki, M. and Zhang, D. (2009), Evaluation of the MEMM, Parameter Estimation and option Pricing for Geometric Lévy Processes, *Asia-Pacific Financial Markets* 16, 111–139.

[43] Fujiwara, T. (2006), From the Minimal Entropy Martingale Measures to the Optimal Strategies for the Exponential Utility Maximization: The Case of Geometric Lévy Processes, *Asia-Pacific Financial Markets* 11, 367–391.

[44] Fujiwara, T. and Miyahara, Y. (2003), The Minimal Entropy Martingale Measures for Geometric Lévy Processes, *Finance and Stochastics* 7, 509–531.

[45] Gerber, H. U. (1979), *An Introduction to Mathematical Risk Theory*, University of Pennsylvania.

[46] Gerber, H. U. and Shiu, E. S. W. (1994), Option Pricing by Esscher Transforms, *Transactions of the Society of Actuaries* XLVI, 99–191.

[47] Goll, T. and Rüschendorf, L. (2001), Minimax and Minimal Distance Martingale Measures and their Relationship to Portfolio Optimization, *Finance and Stochastics* 5(4), 557–581.

[48] Grandits, P. (1999), The p-Optimal Martingale Measure and its Asymptotic Relation with the Minimal Entropy Martingale Measure, *Bernoulli* 5, 225–247.

[49] Grandits, P. and Rheinländer, T. (2002), On the Minimal Entropy Martingale Measure, *The Annals of Probability* 30(3), 1003–1038.

[50] Gundel, A. (2005), Robust Utility Maximization for Complete and Incomplete Market Models, *Finance and Stochastics* 9, 151–176.

[51] He, H. and Pearson, N. D. (1991a), Consumption and Portfolio Policies with Incomplete Markets and Short-Sale Constraints: the Finite-Dimensional Case, *Mathematical Finance* 1, 1–10.

[52] He, H. and Pearson, N. D. (1991b), Consumption and Portfolio Policies with Incomplete Markets and Short-Sale Constraints: the Infinite-Dimensional Case, *Journal of Economic Theory* 54, 259–304.

[53] Hubalek, F. and Sgarra, C. (2006), Esscher Transforms and the Minimal Entropy Martingale Measure for Exponential Lévy Models, *Quantitative Finance* 6, 125–145.

[54] Hurst, S. R., Platen, E. and Rachev, S. T. (1997), Subordinated Markov Index Models: A Comparison, *Financial Engineering and Japanese Markets* 4, 97–124.

[55] Hurst, S. R., Platen, E. and Rachev, S. T. (1999), Option Pricing for a Logstable Asset Price Model, *Math. Comput. Modeling* 29, 105–119.

[56] Ihara, S. (1993), *Information Theory for Continuous System*, World Scientific, Singapore.

[57] Jeanblanc, M., Kloeppel, S. and Miyahara, Y. (2007), Minimal f^q-Martingale Measures for Exponential Lévy Processes, *The Annals of Applied Probability* 17, 1615–1638.

[58] Jeanblanc, M. and Miyahara, Y. (2006), Variance Minimal Martingale Measures for Geometric Lévy Processes, *Discussion Papers in Economics, Nagoya City University* 444, 1–22.

[59] Jeanblanc, M., Yor, M. and Chesney, M. (2009), *Mathematical Methods for Financial Markets*, Springer, London.

[60] Kabanov, Y. M. and Stricker, C. (2002), On the Optimal Portfolio for the Exponential Utility Maximization: Remarks to the Six-Author Paper, *Mathematical Finance* 12, 125–134.

[61] Kallsen, J. (2002), Utility-Based Derivative Pricing in Incomplete Markets, in *Mathematical Finance – Bachelier Congress 2000*, Springer, Berlin, 313–338.

[62] Kallsen, J. and Shiryaev, A. N. (2002), The Cumulant Process and Esscher's Change of Measure, *Finance and Stochastics* 6(4), 397–428.

[63] Karatzas, I., Shreve, S. E. and Xu, G. L. (1991), Martingale and Duality Methods for Utility Maximization in an Incomplete Market, *SIAM Journal on Control and Optimization* 29, 702–730.

[64] Karatzas, I. and Shreve, S. E. (1998), *Methods of Mathematical Finance*, Springer, New York.

[65] Kassberger, S. and Liebmann, T. (2007), *q*-Optimal Martingale Measures for Exponential Lévy Processes (working paper).

[66] Klöppel, S. and Schweizer, M. (2007), Dynamic Utility-based Good Deal Bounds, *Statistics & Decisions* 25, 285–309.

[67] Kohlmann, M. and Niethanner, C. R. (2007), On Convergence to the Exponential Utility Problem, *Stochastic Processes and their Applications* 117, 1813–1834.

[68] Kramkov, D. O. (1996), Optimal Decomposition of Supermartingales and Hedging Contingent Claims in Incomplete Security Markets, *Probability Theory and Related Fields* 105, 459–479.

[69] Kramkov, D. and Schachermayer, W. (1999), The Asymptotic Elasticity of Utility Functions and Optimal Investment in Incomplete Markets, *Annals of Applied Probability* 9, 904–950.

[70] Kunita, H. (2004a), Representation of Martingales with Jumps and Applications to Mathematical Finance, in *Stochastic Analysis and Related Topics in Kyoto, In Honour of Kiyosi Itô, Advanced Studies in Pure Mathematics* 41, 209–232, Mathematical Society of Japan.

[71] Kunita, H. (2004b), Variational Equality and Portfolio Optimization for Price Processes with Jumps, in Akahori J. et al. (eds) *Stochastic Processes and Applications to Mathematical Finance*, 167–194, World Scientific, Singapore.

[72] Kunitomo, N. and Owada, T. (2004), Empirical Liklihood Estimation of Lévy Processes, *CIRJE Discussion Paper, The University of Tokyo* F-272, 1–28.

[73] Kupper, M. and Schachermayer, W. (2008), Representation Results for Law Invariant Time Consistent Functions (preprint).

[74] Madan, D. and Seneta, E. (1990), The Variance Gamma (VG) Model for Share Market Returns, *Journal of Business* 63(4), 511–524.

[75] Madan, D., Carr, P. and Chang, E. (1998), The Variance Gamma Process and option Pricing, *European Finance Review* 2, 79–105.

[76] Maeda, Y., Moriwaki, N. and Miyahara, Y. (2005), On Modeling U.S. Product Liability Risk — An Empirical Analysis, *Working Paper B-5, Center for Risk Research, Shiga University*, 1–20.

[77] Mandelbrot, B. (1963), The Variation of Certain Speculative Prices, *Journal of Business* 36, 394–419.

[78] Merton, R. C. (1976), Option Pricing when Underlying Stock Returns are Discontinuous, *Journal of Financial Economics* 3, 125–144.

[79] Misawa, T. (2010), Simplification of Utility Indifference Net Present Value Method, *OIKONOMIKA* 46(3), Nagoya City University, 123–135.

[80] Mittnik, S. Paolella, M. S. and Rachev, S. T. (1997), A Tail Estimator for the Index of the Stable Paretian Distribution (working paper).

[81] Miyahara, Y. (1996a), Canonical Martingale Measures of Incomplete Assets Markets, in Watanabe S. et al. (eds) *Probability Theory and Mathematical Statistics: Proceedings of the Seventh Japan-Russia Symposium, Tokyo 1995*, 343–352.

[82] Miyahara, Y. (1996b), Canonical Martingale Measures and Minimal Martingale Measures of Incomplete Assets Markets, *The Australian National University Research Report* FMRR 007-96, 95–100.

[83] Miyahara, Y. (1999a), Minimal Entropy Martingale Measures of Jump Type Price Processes in Incomplete Assets Markets, *Asian-Pacific Financial Markets* 6(2), 97–113.

[84] Miyahara, Y. (1999b), Minimal Relative Entropy Martingale Measures and their Applications to Option Pricing Theory, *Proceedings of JIC99, The 5-th JAFEE International Conference*, 316–323.

[85] Miyahara, Y. (2000), Minimal Relative Entropy Martingale Measure of Birth and Death Process, *Discussion Papers in Economics, Nagoya City University* 273, 1–20.

[86] Miyahara, Y. (2001), [Geometric Lévy Process & MEMM] Pricing Model and Related Estimation Problems, *Asia-Pacific Financial Markets* 8(1), 45–60.

[87] Miyahara, Y. (2002), Estimation of Lévy Processes, *Discussion Papers in Economics, Nagoya City University* 318, 1–36.

[88] Miyahara, Y. (2003), *Stock Price Models and Lévy Processes* (in Japanese), Asakura Shoten, Tokyo.

[89] Miyahara, Y. (2004), A Note on Esscher Transformed Martingale Measures for Geometric Lévy Processes, *Discussion Papers in Economics, Nagoya City University* 379, 1–14.

[90] Miyahara, Y. (2006), The [GLP & MEMM] Pricing Model and Related Problems, in Akahori, J. et al. (eds), *Proceedings of the 5th Ritsumeikan International Symposium 'Stochastic Processes and Applications to Mathematical Finance'*, 125–156, World Scientific, Singapore.

[91] Miyahara, Y. (2010), Risk-Sensitive Value Measure Method for Project Evaluation, *Journal of Real Options and Strategy* 3(2), 185–204.

[92] Miyahara, Y. and Moriwaki, N. (2005), Application of [GLP & MEMM] Model to Nikkei 225 Option, *Proceedings of the 7th JAFEE International Conference*, 81–88.

[93] Miyahara, Y. and Moriwaki, N. (2006), Volatility Smile/Smirk Properties of [GLP & MEMM] Models, *RIMS Kokyuroku* 1462, 156–170.

[94] Miyahara, Y. and Moriwaki, N. (2009), Option Pricing Based on Geometric Stable Processes and Minimal Entropy Martingale Measures, in Kijima, M. et al. (eds), *Recent Advances in Financial Engineering: Proceedings of the 2008 Daiwa International Workshop on Financial Engineering*, 119–133, World Scientific, Singapore.

[95] Miyahara, Y. and Novikov, A. (2002), Geometric Lévy Process Pricing Model, *Proceedings of Steklov Mathematical Institute* 237, 176–191.

[96] Miyahara, Y. and Tsujii, Y. (2011), Applications of Risk-Sensitive Value Measure to Portfolio Optimization Problems (preprint).

[97] Niethammer, C. R. (2008), On Convergence to the Exponential Utility Problem with Jumps, *Stochastic Analysis and Applications* 26, 169–196.

[98] Nutz, M. (2009), Power Utility Maximization in Constrained Exponential Lévy Models (preprint).

[99] Overhaus, M., Ferraris, A., Knudsen, T. et al. (2002), *Equity Derivatives: Theory and Applications*, Wiley, Chichester.

[100] Owari, K. (2010), A Note on Utility Maximization with Unbounded Random Endowment, *Asian Pacific Financial Markets* 18(1), 89–103.

[101] Pliska, S. R. (1997), *Introduction to Mathematical Finance*, Blackwell Publishers, Oxford.

[102] Press, S. J. (1972), Estimation in Univariate and Multivariate Stable Distributions, *Journal of the American Statistical Association* 67(340), 842–846.

[103] Rachev, S. and Mittnik, S. (2000), *Stable Paretian Models in Finance*, Wiley, Chichester.

[104] Rheinlander, T and Steiger, G. (2006), The Minimal Entropy Martingale Measure for General Barndorff-Nielsen/Shephard Models 1, *The Annals of Applied Probability* 16(3), 1319–1351.

[105] Rolski, T., Schmidli, H., Schmidt, V. et al. (1999), *Stochastic Processes for Insurance and Finance*, Wiley, Chichester.

[106] Rouge, R. and El Karoui, N. (2000), Pricing Via Utility Maximization and Entropy, *Mathematical Finance* 10(2), 259–276.

[107] Samorodnitsky, G. and Taqqu, M. S. (1994), *Stable Non-Gaussian Random Processes*, Chapman & Hall/CRC, London.

[108] Santacroce, M. (2005), On the Convergence of the p-Optimal Martingale Measures to the Minimal Entropy Martingale Measure, *Stochastic Analysis and Applications* 23, 31–54.

[109] Sato, K. (1999), *Lévy Processes and Infinitely Divisible Distributions*, Cambridge University Press, Cambridge.

[110] Sato, K. (2000), Density Transformation in Lévy Processes, *MaPhySto Lecture Notes* 7.

[111] Schachermayer, W. (2001), Optimal Investment in Incomplete Markets when Wealth may Become Negative, *Annals of Applied Probability* 11, 694–734.

[112] Schoutens, W. (2003), *Lévy Processes in Finance: Pricing Financial Derivatives*, Wiley, Chichester.

[113] Schweizer, M. (1995), On the Minimal Martingale Measure and the Föllmer–Schweizer Decomposition, *Stochastic Analysis and Applications* 13(5), 573–599.

[114] Schweizer, M. (1996), Approximation Pricing and the Variance-Optimal Martingale Measure, *Annals of Probability* 24, 206–236.

[115] Schweizer, M. (2010), Minimal Entropy Martingale Measure, in Cont, R. (ed.), *Encyclopedia of Quantitative Finance*, Wiley, Chichester, 1195–1200.

[116] Shiryaev, A. N. (1999), *Essentials of Stochastic Finance: Facts, Models, Theory*, World Scientific, Singapore.

[117] Shreve, S. E. (2003), *Stochastic Calculus for Finance I: The Binomial Asset Pricing Model*, Springer, New York.

[118] Shreve, S. E. (2004), *Stochastic Calculus for Finance II: Continuous-Time Models*, Springer, New York.

[119] Stricker, C. (2004), Indifference Pricing with Exponential Utility, *Progress in Probability* 58, 323–238.

[120] Williams, D. (1991), *Probability with Martingales*, Cambridge University Press, Cambridge.

[121] Xiao, K., Misawa, T. and Miyahara, Y. (2000), Computer Simulation of [Geometric Lévy Process & MEMM] Pricing Model, *Discussion Papers in Economics, Nagoya City University* 266, 1–16.

[122] Zolotarev, V. M. (1986), *One-dimensional Stable Distributions*, American Mathematical Society, Providence.

Index